D1032928

Theory of Magnetoelectric Properties of 2D Systems

Theory of Magnetoelectric Properties of 2D Systems

S C Chen

Department of Physics, National Cheng Kung University, Tainan, Taiwan 701
Center for Micro/Nano Science and Technology, National Cheng Kung University, Tainan, Taiwan 701

J Y Wu

Center of General Studies, National Kaohsiung Marine University, Kaohsiung 811, Taiwan

C Y Lin

Department of Physics, National Cheng Kung University, Tainan, Taiwan 701

M F Lin

Department of Physics, National Cheng Kung University, Tainan, Taiwan 701

IOP Publishing, Bristol, UK

© IOP Publishing Ltd 2017

All rights reserved. No part of this publication may be reproduced, stored in a retrieval system or transmitted in any form or by any means, electronic, mechanical, photocopying, recording or otherwise, without the prior permission of the publisher, or as expressly permitted by law or under terms agreed with the appropriate rights organization. Multiple copying is permitted in accordance with the terms of licences issued by the Copyright Licensing Agency, the Copyright Clearance Centre and other reproduction rights organisations.

Permission to make use of IOP Publishing content other than as set out above may be sought at permissions@iop.org.

S C Chen, J Y Wu, C Y Lin and M F Lin have asserted their right to be identified as the authors of this work in accordance with sections 77 and 78 of the Copyright, Designs and Patents Act 1988.

ISBN   978-0-7503-1674-3 (ebook)
ISBN   978-0-7503-1672-9 (print)
ISBN   978-0-7503-1673-6 (mobi)

DOI   10.1088/978-0-7503-1674-3

Version: 20171201

IOP Expanding Physics
ISSN 2053-2563 (online)
ISSN 2054-7315 (print)

British Library Cataloguing-in-Publication Data: A catalogue record for this book is available from the British Library.

Published by IOP Publishing, wholly owned by The Institute of Physics, London

IOP Publishing, Temple Circus, Temple Way, Bristol, BS1 6HG, UK

US Office: IOP Publishing, Inc., 190 North Independence Mall West, Suite 601, Philadelphia, PA 19106, USA

Contents

Preface

This book describes a well-developed and up-to-date theoretical model and addresses important advances in diverse quantization phenomena. The key features of this work can be understood from the significant progress of two-dimensional materials, the critical Hamiltonian models, the complete and unusual results, and the clear physical and chemical pictures.

Layered condensed-matter systems, with nano-scaled thickness and unique lattice symmetries are ideal 2D materials for the study of the novel physical, chemical and material phenomena. They have shown very high potential in near-future technological applications. Up to now, emerging 2D materials have included few-layer graphene, silicene, germanene, tinene, phosphorene, MoS_2 and so on. Such systems possess rich intrinsic properties in terms of lattice symmetries, planar or non-planar structures, intra- and inter-layer atomic interactions, multi-orbital bondings, distinct site energies, and spin arrangements. Apparently, composite interactions can create critical Hamiltonians and thus new phenomena.

How to solve the various Hamiltonians is one of the basic topics in physics. In this work, we propose and develop a generalized tight-binding model to comprehend the rich quantization phenomena in 2D materials. The unusual effects, taken into consideration simultaneously, mainly come from the multi-orbital hybridization, the spin-orbital coupling, the intralayer and interlayer atomic interactions, the layer number, the stacking configuration, the site-energy difference, the magnetic field, and the electric field. Especially, the critical Hamiltonians are solved by the exact diagonalization method very efficiently, even for a rather large complex matrix. This model could be further used to investigate electronic properties of 1D and 3D condensed-matter systems. Furthermore, combination with the single- and many-particle theories is very useful in understanding the other essential properties, e.g. optical spectra, transport properties, and Coulomb excitations.

Mainstream 2D materials are predicted to exhibit novel quantization phenomena. There exist three types of quantized Landau levels (LLs) in the sliding graphene systems, the abnormal LL energy spectra in the ABC- and AAB-stacked graphenes, the spin-dominated LLs in (Si, Ge, Sn, MoS_2), two groups of orbital-dependent LLs in tinene, the special dependence of LL energies on quantum number and field strength in few-layer phosphorene, the spin- and valley-dependent LL subgroups in MoS_2, and the frequent LL splittings, crossings, and anti-crossings under a composite electric and magnetic field.

Clear physical and chemical pictures are proposed to explain the diverse quantization. The magnetic quantization is greatly enriched by the various mechanisms. The structure transformation between two high-symmetry stacking configurations can create new types of LLs. The unusual band structures are reflected in the irregular LL energies. The LL degeneracy is reduced by the destruction of inversion symmetry, the absence of constant-energy multi-valleys, or the site-energy differences of different orbitals. The orbital-dependent LLs come from the multi-orbital hybridization. The spin-orbital coupling is responsible for the spin-dominated configuration and even for the spin-split LLs. The valley-dependent LLs are induced by the cooperation of distinct site energies and magnetic field. Moreover, the interlayer atomic interactions and the electric field will result in LL anti-crossing behavior.

This book shows that the generalized tight-binding model has been successfully developed to explore the diverse quantization in emerging 2D materials. This model can deal with the structure-, orbital-, spin- and field-dependent Hamiltonians. The studies on the critical Hamiltonians could attract much attention from researchers in the scientific community, not only for the study of 2D systems, but also for the exploration of other 1D & 3D systems.

IOP Publishing

Theory of Magnetoelectric Properties of 2D Systems

S C Chen, J Y Wu, C Y Lin and M F Lin

Chapter 1

Introduction

How to solve the Hamiltonian is one of the basic topics in the science of physics. It is very interesting to comprehend the diverse quantization phenomena arising from the various Hamiltonians in condensed-matter systems, especially for the feature-rich magnetic quantization. Such Hamiltonians possess complex effects coming from the multi-orbital chemical bondings, spin-orbital coupling (SOC), magnetic field ($\mathbf{B} = B_z \hat{z}$), electric field ($\mathbf{E} = E_z \hat{z}$), interlayer hopping integrals, number of layers, stacking configurations, curved surfaces, hybridized structures, and distinct dimensionalities. The generalized tight-binding model, based on the subenvelope functions of different sublattices, is developed to include all the critical interactions simultaneously [1–12]. The magnetically quantized energy spectra and wave functions can be computed very efficiently by the exact diagonalization method, even for a rather large Hamiltonian matrix with complex elements. This model has been used to make systematic studies on three-dimensional (3D) graphites [1–4], 2D graphenes [5–10], 1D graphene nanoribbons, [11, 12] and carbon nanotubes [13]. It is further extended to mainstream layered materials, e.g. other group-IV systems [14–17], and MoS_2 [18–20]. Moreover, the generalized tight-binding model can directly combine with single- and many-particle theories to study other essential physical properties, such as magneto-optical properties [12, 21–26], Coulomb excitations [15, 27, 28], and quantum transport [29].

However, the perturbation method is frequently used to investigate low-energy electronic states and magnetic quantization. It is suitable for condensed-matter systems with simple and monotonous band structures. For example, the effective-mass approximation can deal with the magnetic quantization in monolayer graphene [30–33], AA- and AB-stacked few-layer graphenes [24–29], monolayer silicene and germanene [34], MoS_2 [35, 36], and phosphorene [43]. The low-energy perturbation at the symmetry point will become too complex or cumbersome to magnetically quantize the multi-valley and/or multi-orbital electronic states, such as the magnetic quantization for the oscillatory energy bands in ABC- and AAB-stacked graphenes [7, 10], the seriously distorted Dirac-cone structure in sliding bilayer graphenes [5], the three constant-energy loops due to the significant sp^3 bondings in monolayer tinene [13], and the mixed energy bands in hybridized carbon systems [12, 44, 45]. Furthermore, it is very difficult to resolve the complex quantization phenomena in the presence of non-uniform or composite external fields [16–18].

Layered condensed-matter systems have attracted a lot of experimental [49–54] and theoretical studies [5–10, 24], mainly owing to their nano-scaled thickness and specific symmetries. They are ideal 2D materials for studying novel physical, chemical and material phenomena. Furthermore, such systems have shown a high potential for future technological applications, e.g. nano-electronics [55–62], optoelectronics [63–74], and energy storage [75–87]. Few-layer graphenes have been successfully synthesized by distinct experimental methods, such as mechanical exfoliation [88–98], electrostatic manipulation using scanning tunneling microscopy (STM) [15, 16, 99, 102, 103], and chemical vapor deposition [47, 104–122]. Four kinds of typical stacking configurations, AAA [108, 122], ABA [107, 110, 113], ABC [105, 106] and AAB [47, 107], are clearly identified in experimental measurements. It should be noted that a STM tip can generate continuous changes in the stacking configuration, e.g. configuration transformations among the ABA, ABC and AAB stackings [15].

The essential electronic properties of planar graphenes are dominated by the $2p_z$-orbital hybridization, the hexagonal honeycomb symmetry (figure 1(a)), the stacking configuration and the number of layers. The main features of low-lying energy bands are further reflected in the rich magnetic quantization. Specially, the tri-layer AAA, ABA, ABC and AAB stackings, respectively, possess unusual energy dispersions, namely (1) linearly intersecting bands (the almost isotropic Dirac-cone structures) [9, 24], (2) parabolic bands and linear bands [8], (3) weakly dispersive bands, sombrero-shaped bands, and linear bands [7]; (4) oscillatory bands, sombrero-shaped bands, and parabolic bands [10]. The neighboring electronic states in the presence of a uniform perpendicular magnetic field are further quantized into Landau levels (LLs), with the high degeneracy and oscillatory spatial probability distributions (discussed in chapter 3). The typical stacking systems exhibit novel LLs, in which the rich magneto-electronic properties include diverse B_z-dependent energy spectra [7, 9], asymmetric energy spectra at the Fermi level (E_F) [7–9], non-crossing, crossing and anti-crossing behaviors, main and side modes [9], and configuration- and E_z-created splitting states [8, 10, 36]. As for the sliding bilayer graphenes, the configuration transformation between AA and AB stackings will create the thorough destruction of the Dirac-cone structures [5, 126]. Three types of LLs, namely well-behaved, perturbed and undefined LLs, are predicted to be revealed in the changes from the linear to the parabolic bands [5, 126].

Few-layer germanene and silicene can be synthesized on distinct substrate surfaces, e.g. Si on Ag(111), Ir(111) and ZrB_2 surfaces [127–131]; Ge on Pt(111), Au(111) and Al(111) surfaces [132–134]. Germanene and silicene possess buckled structures with a mixed sp^2-sp^3 bonding rather than a sp^2 bonding, since the relatively weak chemical bonding between the larger atoms cannot maintain a planar structure (figure 1(b)). These two systems have significant SOCs that are

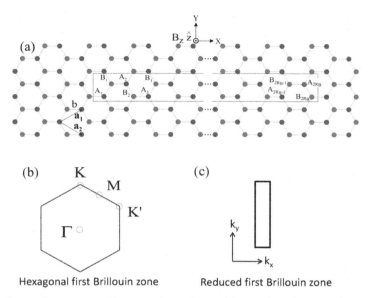

Figure 1. (a) Geometric structure of honeycomb graphene with an enlarged rectangular unit cell in $B_z\hat{z}$. (b) The first Brillouin zone of the hexagonal lattice and the high-symmetry points (red circles), and (c) the reduced first Brillouin zone. $\mathbf{a_1}$ and $\mathbf{a_2}$ in (a) are lattice vectors, and the subscript of A_l corresponds to the lth atom.

much stronger than that in graphene. The SOCs can separate the Dirac-cone structures built from the dominating $3p_z$ or $4p_z$ orbitals; that is, the intrinsic systems are narrow-gap semiconductors ($E_g \sim 45$ meV for Ge and $E_g \sim 5$ meV for Si) [1, 136]. Furthermore, the application of a uniform perpendicular electric field leads to the modulation of the energy gap and the splitting of spin-related configurations [34, 137, 138]. The magneto-electronic properties are greatly enriched by SOC and E_z, including the modified B_z-dependent energy spectra, the spin-up- and spin-down-dominated states, and the E_z-generated crossing and anti-crossing behaviors (results in chapter 4).

Monolayer tinene is successfully fabricated on a substrate of bismuth telluride [14], while a monolayer Pb system is absent in experimental measurements up to now. Theoretical studies show that the single-layer Sn and Pb systems have rather strong sp^3 bondings and SOCs [1, 140]. Apparently, the complex chemical bondings from four (s,p_x,p_y,p_z) orbitals need to be included in low-energy model calculations. However, the low-lying electronic structures of graphene, silicene and germanene are mainly determined by the p_z orbitals. The very pronounced mixing effects of multi-orbital bondings and SOCs can create the p_z- and (p_x,p_y)-dominated energy bands near E_F, indicating the existence of multi-constant-energy loops. There exist two groups of low-lying LLs, with different orbital components, spin configurations, localization centers, state degeneracy, and B_z- and E_z-dependencies [13]. Specifically, the LL splitting and anti-crossing behaviors strongly depend on the type of orbitals and the external fields. The competitive or cooperative relations among the orbital hybridizations, SOC, **B** and **E** are worthy of detailed investigations (chapter 4).

Group-V phosphorus possesses several allotropes in which black phosphorus (BP) is the most stable phase under normal experimental conditions [141]. Few-layer phosphorene is successfully obtained using the mechanical cleavage approach [9, 142], liquid exfoliation [144–146], and mineralizer-assisted short-way transport reactions [147–149]. Experimental measurements show that the BP-based field effect transistor has an on/off ratio of 10^5 and a carrier mobility at room temperature as high as 10^3 cm^2 (Vs)$^{-1}$. BP is expected to play an important role in next-generation electronic devices [9, 142]. Phosphorene exhibits a puckered structure related to the sp^3 hybridization of four (3s,3p$_x$,3p$_y$,3p$_z$) orbitals. The deformed hexagonal lattice of monolayer BP has four atoms [1], while the group-IV honeycomb lattice includes two atoms. The low-lying energy bands are highly anisotropic, e.g. the linear and parabolic dispersions near E_F, respectively, along the $\widehat{k}_x$ and $\widehat{k}_y$ directions. The anisotropic behavior is clearly reflected in other physical properties, as verified by recent measurements of optical spectra and transport properties [142, 151]. BP has a middle energy gap of ~1.5–2eV at the Γ point, being quite different from the narrow or zero gaps of group-IV systems. The low-lying energy dispersions, which are dominated by 3p$_z$ orbitals, can be described by a four-band model with complicated multi-hopping integrals [1]. The low-energy electronic structure is easily tuned by a perpendicular electric field, e.g. the monotonic increase of E_g with E_z in monolayer BP, and the transition from a semiconducting to a gapless system in bilayer BP [152, 153]. In sharp contrast with the group-IV monolayer systems, monolayer phosphorene presents unique LLs with an asymmetric energy spectrum about E_F, a reduced state degeneracy, and a spin-independent configuration (results in chapter 5). The important differences mainly arise from the geometric structure, the orbital hybridization, and the SOC. The magnetic quantization is greatly diversified by the number of layers.

Transition metal dichalcogenide monolayers can be produced by micromechanical cleavage [154–157], liquid-phase exfoliation [158, 159] and chemical vapor deposition [160–164]. Due to the unusual electronic and optical properties, various technological applications have been proposed for these materials, such as electronic [49, 165–168] and optoelectronic [168–170] devices. The high potential of field-effect transistors is supported by the room-temperature carrier mobility of over 200 cm^2 (Vs)$^{-1}$ and the high on/off ratio of ~10^8 [49]. Furthermore, experimental measurements show that they have a direct band gap in the visible frequency range [157, 171, 172] and valley-dependent optical selection rules [169, 173]. The stronger SOC and the inversion symmetry breaking lead to the spin- and valley-dependent electronic states [169]. The MoS$_2$-related systems are very suitable for investigating spintronics and valleytronics. Specifically, the single-layer MoS$_2$ is composed of staggered honeycomb-like lattice structures in which a single layer of Mo atoms is sandwiched between two sulfur layers. This semiconducting system has a direct energy gap of ~1.59 eV [15, 26]. The low-lying electronic states near three valleys centered at the (K,K$'$) and Γ points (figure 1(d)) are dominated by the (4d$_{z^2}$,4d$_{xy}$,4d$_{x^2-y^2}$) orbitals of the Mo atoms. The SOC can effectively destroy the spin degeneracy of the energy bands, especially for the valence one contributed by the 4d$_{xy}$ and 4d$_{x^2-y^2}$ orbitals. The quantized LLs are characterized by the dominating orbitals and spin configurations [15], being enriched by the constant-energy loops in three valleys. The degeneracy of the K and K' valleys is further lifted by **B**, owing to the cooperation of the site-energy difference and the magnetic quantization (results in chapter 6).

As for the electronic energy spectra of valence and conduction states, the theoretical predictions could be directly verified from scanning tunneling spectroscopy (STS). The STS measurements, in which the tunneling conductance (dI/dV) is proportional to the density of states (DOS), could serve as efficient methods to examine the special structures of DOS arising from the van Hove singularities. They have been successfully utilized to verify the various electronic properties in graphene-related systems, such as few-layer graphenes [6, 47, 60, 176–179], graphite [4, 182], graphene nanoribbons [183–185], and carbon nanotubes [48, 186]. As a result of the discrete characteristic of LLs, each prominent peak in DOS is easily identified from experimental measurements. Up to now, the diverse B_z-dependent LL energy spectra are confirmed for the AB-stacked few-layer graphenes [64–66] and graphite [67, 68]; discussions in the final paragraph of chapter 6. According to the symmetries of spatial probability distributions, the quantized LLs will create the selection rules of magneto-optical excitations. Both infrared transmission [18–20] and magneto-Raman spectroscopies [21] are powerful tools in the identification of effective excitation channels. A specific selection rule of $(\Delta n)_s = \pm 1$ is verified to be present in the AB-stacked few-layer graphenes [18] and graphite [19, 20]; the final paragraph of chapter 3).

Specifically, non-uniform magnetic quantization is one of the diverse phenomena under the external fields. A spatially modulated magnetic field could be achieved by an array of superconducting or ferromagnetic stripes [197–199]. The 2D free electron gases formed by GaAs/AlGaAs heterojunctions in modulated magnetic fields are well-established systems and have been subjects of active studies in the past two decades [197–201]. However, the systems for graphene under such fields are not yet realized experimentally. There are many theoretical calculations, for which most are focused on the electronic [2, 4, 18], optical [3, 16], and transport properties [9]. An inhomogenous magnetic field in monolayer graphene is predicted to create some interesting and distinct properties, such as the anisotropy of the low-lying energy bands, creation of quasi-Landau levels (QLLs), coexistence of the well-known and extra magneto-optical selection rule, and quantum Hall effects. The calculated results reflect the fact that this field is able to flock the neighboring electronic states together, while the highly degenerate LLs hardly survive under non-uniform magnetic quantization. Moreover, a composite magnetic field, with uniform and modulated components, could be utilized to tune the main characteristics of LLs [4, 18]. The complex competition/cooperation between these two components will determine the diversified properties.

Few-layer graphene, silicene, germanene, tinene, phosphorene and MoS_2 are chosen for a systematic study on the essential electronic properties. The dependences of electronic structures, wave functions and DOSs on the distinct sublattices, atomic orbitals, spin configurations, and electric and magnetic fields are explored in detail. The physical and chemical pictures are proposed to explain the diverse phenomena. The contents of this work are organized as follows. Chapter 2 is focused on the direct derivations of the various Hamiltonians without/with the external fields. According to the generalized tight-binding model, the independent zero-field and magnetic Hamiltonian matrix elements are presented in analytic forms, covering the intralayer and interlayer hopping integrals, multi-orbital atomic hybridizations, spin-orbital couplings, and site energies. How to diagonalize a huge Hermitian matrix very efficiently is explored in detail. The feature-rich electronic properties, being characterized by energy dispersions, band gaps, field-dependent LL energy spectra, spatial distributions of LL wave functions and special structures in DOS, are discussed thoroughly in chapters 3–6. The LL anti-crossing behavior, valley splitting and spin dominance, which are clearly revealed in the B_z- and E_z-dependent energy spectra, are worthy of detailed investigations. Comparisons between the theoretical predictions and the recent experimental measurements are also made. Specifically, non-uniform magnetic quantization is the focus of chapter 7, covering the main features of QLLs, the important differences between QLLs and LLs, and the strong effects of a modulated magnetic field on the LL characteristics. Concluding remarks, as given in chapter 8, address the important progress in understanding the emergent layered materials and further development by the consistent combination of this model with other theories.

References

[1] Ho Y H, Wang J, Chiu Y H, Lin M F and Su W P 2011 Characterization of Landau subbands in graphite: a tight-binding study *Phys. Rev.* B **83** 121201

[2] Ho C H, Ho Y H, Liao Y Y, Chiu Y H, Chang C P and Lin M F 2012 Diagonalization of Landau level spectra in rhombohedral graphite *J. Phys. Soc. Japan* **81** 024701

[3] Chen R B, Chiu C W and Lin M F 2015 Magnetoplasmons in simple hexagonal graphite *RSC Adv.* **5** 53736–40

[4] Ho C H, Chang C P and Lin M F 2014 Landau subband wave functions and chirality manifestation in rhombohedral graphite *Solid State Commun.* **197** 11–5

[5] Huang Y K, Chen S C, Ho Y H, Lin C Y and Lin M F 2014 Feature-rich magnetic quantization in sliding bilayer graphenes *Sci. Rep.* **4** 7509

[6] Ho J H, Lai Y H, Chiu Y H and Lin M F 2008 Landau levels in graphene *Physica* E **40** 1722–5

[7] Lin C Y, Wu J Y, Chiu Y H and Lin M F 2014 Stacking-dependent magneto-electronic properties in multilayer graphenes *Phys. Rev.* B **90** 205434

[8] Lai Y H, Ho J H, Chang C P and Lin M F 2008 Magnetoelectronic properties of bilayer Bernal graphene *Phys. Rev.* B **77** 085426

[9] Lin C Y, Wu J Y, Ou Y J, Chiu Y H and Lin M F 2015 Magneto-electronic properties of multilayer graphenes *Phys. Chem. Chem. Phys.* **17** 26008–35

[10] Do T N, Lin C Y, Lin Y P, Shih P H and Lin M F 2015 Configuration-enriched magnetoelectronic spectra of AAB-stacked trilayer graphene *Carbon* **94** 619–32

[11] Huang Y C, Chang C P and Lin M F 2007 Magnetic and quantum confinement effects on electronic and optical properties of graphene ribbons *Nanotechnology* **18** 495401

[12] Chung H C, Chang C P, Lin C Y and Lin M F 2016 Electronic and optical properties of graphene nanoribbons in external fields *Phys. Chem. Chem. Phys.* **18** 7573–616

[13] Shyu F L, Chang C P, Chen R B, Chiu C W and Lin M F 2003 Magnetoelectronic and optical properties of carbon nanotubes *Phys. Rev.* B **67** 045405

[14] Chen S C, Wu C L, Wu J Y and Lin M F 2016 Magnetic quantization of sp^3 bonding in monolayer gray tin *Phys. Rev.* B **94** 045410

[15] Wu J Y, Lin C Y, Gumbs G and Lin M F 2015 The effect of perpendicular electric field on temperature-induced plasmon excitations for intrinsic silicene *RSC Adv.* **5** 51912–18

[16] Wu J Y, Chen S C and Lin M F 2014 Temperature-dependent Coulomb excitations in silicene *New J. Phys.* **16** 125002

[17] Wu J Y, Chen S C, Gumbs G and Lin M F 2016 Feature-rich electronic excitations in external fields of 2D silicene *Phys. Rev.* B **94** 205427

[18] Ho Y H, Su W P and Lin M F 2015 Hofstadter spectra for d-orbital electrons: a case study on MoS_2 *RSC Adv.* **5** 20858–64

[19] Ho Y H, Chiu C W, Su W P and Lin M F 2014 Magneto-optical spectra of transition metal dichalcogenides: a comparative study *Appl. Phys. Lett.* **105** 222411

[20] Ho Y H, Wang Y H and Chen H Y 2014 Magnetoelectronic and optical properties of a MoS_2 monolayer *Phys. Rev.* B **89** 55316

[21] Huang Y C, Chang C P and Lin M F 2008 Magnetoabsorption spectra of bilayer graphene ribbons with Bernal stacking *Phys. Rev.* B **78** 115422

[22] Lin Y P, Lin C Y, Ho Y H, Do T N and Lin M F 2015 Magneto-optical properties of ABC-stacked trilayer graphene *Phys. Chem. Chem. Phys.* **17** 15921–7

[23] Chen R B, Chiu Y H and Lin M F 2014 Beating oscillations of magneto-optical spectra in simple hexagonal graphite *Comput. Phys. Commun.* **189** 60–5

[24] Chen R B, Chiu Y H and Lin M F 2012 A theoretical evaluation of the magneto-optical properties of AA-stacked graphite *Carbon* **54** 268–76

[25] Ho Y H, Wu J Y, Chen R B, Chiu Y H and Lin M F 2010 Optical transitions between Landau levels: AA-stacked bilayer graphene *Appl. Phys. Lett.* **97** 101905

[26] Ho Y H, Chiu Y H, Lin D H, Chang C P and Lin M F 2010 Magneto-optical selection rules in bilayer Bernal graphene *ACS Nano* **4** 1465–72

[27] Wu J Y, Gumbs Godfrey and Lin M F 2014 Combined effect of stacking and magnetic field on plasmon excitations in bilayer graphene *Phys. Rev.* B **89** 165407

[28] Wu J Y, Chen S C, Roslyak Oleksiy, Gumbs Godfrey and Lin M F 2011 Plasma excitations in graphene: their spectral intensity and temperature dependence in magnetic field *ACS Nano.* **5** 1026–32

[29] Do T N, Chang C P, Shih P H and Lin M F 2017 Stacking-enriched magneto-transport properties of few-layer graphenes *Phys. Chem. Chem. Phys.* **19** 29525

[30] Zheng Y and Ando T 2002 Hall conductivity of a two-dimensional graphite system *Phys. Rev.* B **65** 245420

[31] Sharapov S G, Gusynin V P and Beck H 2004 Magnetic oscillations in planar systems with the Dirac-like spectrum of quasiparticle excitations *Phys. Rev.* B **69** 075104

[32] Gusynin V P and Sharapov S G 2005 Unconventional integer quantum Hall effect in graphene *Phys. Rev. Lett.* **95** 146801

[33] Goerbig M O 2011 Electronic properties of graphene in a strong magnetic field *Rev Mod. Phys.* **83** 1193–243

[34] Chang C P 2011 Exact solution of the spectrum and magneto-optics of multilayer hexagonal graphene *J. Appl. Phys.* **110** 013725

[35] McCann E and Fal'ko V I 2006 Landau-level degeneracy and quantum Hall effect in a graphite bilayer *Phys. Rev. Lett.* **96** 086805

[36] Koshino M and McCann E 2011 Landau level spectra and the quantum Hall effect of multilayer graphene *Phys. Rev.* B **83** 165443

[37] Min H and MacDonald A H 2008 Chiral decomposition in the electronic structure of graphene multilayers *Phys. Rev.* B **77** 155416

[38] Sena S H R, Pereira J M Jr., Peeters F M and Farias G A 2011 Landau levels in asymmetric graphene trilayers *Phys. Rev.* B **84** 205448

[39] McCann E and Koshino M 2013 The electronic properties of bilayer graphene *Rep. Prog. Phys.* **76** 056503

[40] Ezawa M 2012 Valley-polarized metals and quantum anomalous Hall effect in silicene *Phys. Rev. Lett.* **109** 055502

[41] Tahir M, Vasilopoulos P and Peeters F M 2016 Quantum magnetotransport properties of a MoS$_2$ monolayer *Phys. Rev.* B **93** 035406

[42] Kormányos Andor, Rakyta Péter and Burkard Guido 2015 Landau levels and Shubnikov-de Haas oscillations in monolayer transition metal dichalcogenide semiconductors *New J. Phys.* **17** 103006

[43] Rodin A S, Carvalho A and Castro Neto A H 2014 Strain-induced gap modification in black phosphorus *Phys. Rev. Lett.* **112** 176801

[44] Li T S, Chang S C, Lien J Y and Lin M F 2008 Electronic properties of nanotube-ribbon hybrid systems *Nanotechnology* **19** 105703

[45] Li T S, Lin M F and Chang S C 2010 Quantum conductance in nanotube-ribbon hybrids *J. Appl. Phys.* **107** 063714

[46] Ou Y C, Chiu Y H, Yang P H and Lin M F 2014 The selection rule of graphene in a composite magnetic field *Optics Express* **22** 7473

[47] Ou Y C, Chiu Y H, Lu J M, Su W P and Lin M F 2013 Electric modulation effect on magneto-optical spectrum of monolayer graphene *Comput. Phys. Commun.* **184** 1821–6

[48] Ou Y C, Sheu J K, Chiu Y H, Chen R B and Lin M F 2011 Influence of modulated fields on the Landau level properties of graphene *Phys. Rev.* B **83** 195405

[49] Radisavljevic B, Radenovic A, Brivio J, Giacometti V and Kis A 2011 Single-layer MoS$_2$ transistors *Nature Nanotech* **6** 147–50

[50] Hao K, Moody G, Wu F, Dass C, Xu L, Chen C H, Li M Y, Li L J, MacDonald A and Li X 2016 Direct measurement of exciton valley coherence in monolayer WSe$_2$ *Nature Phys* **12** 677–82

[51] Zhang C, Chen Y, Huang J K, Wu X, Li L J, Yao W, Tersoff J and Shih C K 2016 Visualizing band offsets and edge states in bilayer–monlayer transition metal dichalcogenides lateral heterojunction *Nat. Commun.* **7** 10349

[52] Li H, Shi Y, Chiu M H and Li L J 2015 Emerging energy applications of two-dimensional layered transition metal dichalcogenides *Nano Energy* **18** 293–305

[53] Hsu W T, Chen Y L, Chen C H, Liu P S, Hou T H, Li L J and Chang W H 2015 Optically initialized robust valley-polarized holds in monolayer WSe$_2$ *Nature Comm* **6** 8963

[54] Qi J, Lan Y W, Stieg A, Chen J H, Zhong Y L, Li L J, Chen C D, Zhang Y and Wang K 2015 Piezoelectric effect in CVD-grown atomic-monolayer triangular MoS$_2$ piezotronics *Nat. Commun.* **6** 7430

[55] Engel M, Steiner M, Lombardo A, Ferrari A C, v. Lohneysen H and Avouris P *et al* 2012 Light matter interaction in a micro-cavity controlled graphene transistor *Nat. Commun.* **3** 906–11

[56] Khan F, Baek S H and Kim J H 2016 One-step and controllable bipolar doping of reduced graphene oxide using TMAH as reducing agent and doping source for field effect transistors *Carbon* **100** 608–16

[57] Kumar A, Tyagi P, Dagar J and Srivastava R 2016 Tunable field effect properties in solid state and flexible graphene electronics on composite high-low k dielectric *Carbon* **99** 579–84

[58] Li M Y, Shi Y, Cheng C C, Lu L S, Lin Y C and Tang H L *et al* 2015 Epitaxial growth of a monolayer WSe_2-MoS_2 lateral p-n junction with an atomically sharp interface *Science* **349** 524–58

[59] Wang Y-T, Luo C-W, Yabushita A, Wu K H, Kobayashi T, Chen C-H and Li L-J 2015 Ultrafast multi-level logic gates with spin-valley coupled polarization anisotropy in monolayer MoS_2 *Sci. Rep.* **5** 8289

[60] Kasry A, Kuroda M A, Martyna G J, Tulevski G S and Bol A A 2010 Chemical doping on large-area stacked graphene films for use as transparent conducting electrodes *ACS Nano* **4** 3839–44

[61] Xiang Q, Yu J and Jaroniec M 2012 Graphene-based semiconductor photocatalysts *Chem. Soc. Rev.* **41** 782–96

[62] Rechtsman M C, Zeuner J M, Plotnik Y, Lumer Y, Podolsky D, Dreisow F, Nolte S, Segev M and Szameit A 2013 Photonic Floquet topological insulators *Nature* **496** 196–200

[63] Koppens F H L, Mueller T, Avouris Ph, Ferrari A C, Vitiello M S and Polini M 2014 Photodetectors based on graphene, other two-dimensional materials and hybrid systems *Nat. Nanotechnol.* **9** 780

[64] Bonaccorso F, Sun Z, Hasan T and Ferrari A C 2010 Graphene photonics and optoelectronics *Nat. Photonics* **4** 611

[65] Tseng S F, Haiso W T, Cheng P Y, Chung C K, Lin Y S and Chien S C *et al* 2016 Graphene-based chips fabricated by ultraviolet laser patterning for an electrochemical impedance spectroscopy *Sensors Actuators* B **226** 342

[66] Liu J B, Li P J, Chen Y F, Song X B, Mao Q and Wu Y *et al* 2016 Flexible terahertz modulator based on coplanar-gate graphene field-effect transistor structure *Opt. Lett.* **41** 816

[67] Deng H Y, Chen X F, Malomed B A, Panoiu N C and Ye F W 2016 Tunability and robustness of Dirac points of photonic nanostructures *IEEE J. Sel. Top. Quantum Electron* **22** 5000509

[68] Yan H G, Li X S, Chandra B, Tulevski G, Wu Y G, Freitag M, Zhu W J, Avouris P and Xia F G 2012 Tunable infrared plasmonic devices using graphene/insulator stacks *Nat. Nanotechnol.* **7** 330

[69] Kocaman S, Aras M S, Hsieh P, McMillan J F, Biris C G, Panoiu N C, Yu M B, Kwong D L, Stein A and Wong C W 2011 Zero phase delay in negative-refractive-index photonic crystal superlattices *Nat. Photonics* **5** 499

[70] Tassin P, Koschny T and Soukoulis C M 2013 Graphene for terahertz applications *Science* **341** 620

[71] Vakil A and Engheta N 2011 Transformation optics using graphene *Science* **332** 1291

[72] Deng H Y, Ye F W, Malomed B A, Chen X F and Panoiu N C 2015 Spin-dependent refraction at the atomic step of transition-metal dichalcogenides *Phys. Rev.* B **91** 201402

[73] Deng H Y, Chen X F, Malomed B A, Panoiu N C and Ye F W 2015 Transverse Anderson localization of light near Dirac points of photonic nanostructures *Sci. Rep.* **5** 15585

[74] de Abajo G and Javier F 2014 Graphene plasmonics: challenges and opportunities *ACS Photonics* **1** 135

[75] Shown I, Hsu H C, Chang Y, Lin C H, Roy P K and Ganguly A *et al* 2014 Highly efficient visible light photocatalytic reduction of CO^2 to hydrocarbon fuels by Cu-nanoparticle decorated graphene oxide *Nano Lett.* **14** 6097

[76] Hsu H C, Shown I, Wei H Y, Chang Y C, Du H Y and Lin Y G *et al* 2013 Graphene oxide as a promising photocatalyst for CO_2 to methanol conversion *Nanoscale* **5** 262

[77] Ibram G 2016 Electrochemical conversion of carbon dioxide into renewable fuel chemicals—The role of nanomaterials and the commercialization *Renew. Sust. Energ. Rev* **59** 1269

[78] Baughman R H, Zakhidov A A and de Heer W A 2002 Carbon nanotubes—the route toward applications *Science* **297** 787

[79] Simon P and Gogotsi Y 2008 Materials for electrochemical capacitors *Nat. Mat* **7** 845

[80] Stoller M D, Park S J, Zhu Y W, An J H and Ruoff R S 2008 Graphene based ultracapacitors *Nano Lett.* **8** 3498–502

[81] Chan C K, Peng H L, Liu G, McIlwrath K, Zhang X F, Huggins R A and Cui Y 2008 High-performance lithium battery anodes using silicon nanowires *Nat. Nat Nanotechnol* **3** 31

[82] Wang G P, Zhang L and Zhang J J 2012 A review of electrode materials for electrochemical supercapacitors *Chem. Soc. Rev.* **41** 797

[83] Bulusheva L G, Arkhipov V E, Fedorovskaya E O, Zhang S, Kurenya A G, Kanygin M A, Asanov I P, Tsygankova A R, Chen X H and Song H H 2016 Fabrication of free-standing aligned multiwalled carbon nanotube array for Li-ion batteries *J. Power Sources* **311** 42

[84] Bissett M A, Worrall S D, Kinloch I A and Dryfe R A W 2016 Comparison of two-dimensional transition metal dichalcogenides for electrochemical supercapacitors *Electrochim. Acta* **201** 30

[85] Li J Y, Wang L and He X M 2016 Phosphorus-based composite anode materials for secondary batteries *Prog. Chem.* **28** 193

[86] Sun Y G, Wu Q and Shi G Q 2011 Graphene based new energy materials *Energy Environ. Sci.* **4** 1113

[87] Gwon H, Kim H S, Lee K U, Seo D H, Park Y C, Lee Y S, Ahn B T and Kang K 2011 Flexible energy storage devices based on graphene paper *Energy Environ. Sci.* **4** 1277

[88] Hattendorf S, Georgi A, Liebmann M and Morgenstern M 2013 Networks of ABA and ABC stacked graphene on mica observed by scanning tunneling microscopy *Surf. Sci.* **610** 53

[89] Novoselov K S, Geim A K, Morozov S V, Jiang D, Zhang Y, Dubonos S V and Firsov A A 2004 Electric field effect in atomically thin carbon films *Science* **306** 666

[90] Jayasena B and Subbiah S 2011 A novel mechanical cleavage method for synthesizing few-layer graphenes *Nanoscale Res. Lett.* **6** 95

[91] Cooper A J, Wilson N R, Kinloch I A and Dryfe R A W 2014 Single stage electrochemical exfoliation method for the production of few-layer graphene via intercalation of tetraalky-lammonium cations *Carbon* **66** 340

[92] Noroozi M, Zakaria A, Radiman S and Wahab Z A 2016 Environmental synthesis of few layers graphene sheets using ultrasonic exfoliation with enhanced electrical and thermal properties *PLOS One* **11** e0152699

[93] Dobbelin M, Ciesielski A, Haar S, Osella S, Bruna M and Minoia A *et al* 2016 Light-enhanced liquid-phase exfoliation and current photoswitching in graphene-zobenzene composites *Nat. Commun.* **7** 11090

[94] Majee S, Song M, Zhang S L and Zhang Z B 2016 Scalable inkjet printing of shear-exfoliated graphene transparent conductive films *Carbon* **102** 51

[95] Arao Y, Mizuno Y, Araki K and Kubouchi M 2016 Mass production of high-aspect-ratio few-layer-graphene by high-speed laminar flow *Carbon* **102** 330

[96] Bracamonte M V, Lacconi G I, Urreta S E and Torres L E F Foa 2014 On the nature of defects in liquid-phase exfoliated graphene *J. Phys. Chem.* C **118** 15455

[97] Song P, Zhang X, Sun M, Cui X and Lin Y 2012 Synthesis of graphene nanosheets via oxalic acid-induced chemical reduction of exfoliated graphite oxide *RSC Adv.* **2** 1158 .

[98] Dou L, Cui F, Yu Y, Khanarian G, Eaton S W, Yang Q, Resasco J, Schildknecht C, Schierle-Arndt K and Yang P 2016 Solution-processed copper/reduced-graphene-oxide core/shell nanowire transport conductors *ACS Nano* **10** 2600

[99] Yin L J, Wang W X, Feng K K, Nie J C, Xiong C M and Dou R F *et al* 2015 Liquid-assisted tip manipulation: fabrication of twisted bilayer graphene superlattices on HOPG *Nanoscale* **7** 14865

[100] Xu P, Ackerman M L, Barber S D, Schoelz J K, Qi D J and Thibado P M *et al* 2013 Graphene manipulation on 4H-SiC(0001) using scanning tunneling microscopy *Jcpan J. Appl. Phys.* **52** 035104 .

[101] Xu P, Yang Y R, Qi D, Barber S D, Schoelz J K and Ackerman M L *et al* 2012 Electronic transition from graphite to graphene via controlled movement of the top layer with scanning tunneling microscopy *Phys. Rev.* B **86** 085428

[102] Kurys Y I, Ustavytska O O, Koshechko V G and Pokhodenko V D 2016 Structure and electrochemical properties of multilayer graphene prepared by electrochemical exfoliation of graphite in the presence of benzoate ions *RSC Adv.* **6** 36050

[103] Xu P, Ackerman M L, Barber S D, Schoelz J K, Thibado P M and Wheeler V D *et al* 2013 Competing scanning tunneling microscope tip-interlayer interactions for twisted multilayer graphene on the a-plane SiC surface *Sur. Sci.* **617** 113

[104] Ye S and Oh W C 2016 Demonstration of enhanced the photocatalytic effect with PtSe$_2$ and TiO$_2$ treated large area graphene obtained by CVD method *Mat. Sci. Semicon. Proc.* **48** 106

[105] Norimatsu W and Kusunoki M 2010 Selective formation of ABC-stacked graphene layers on SiC(0001) *Phys. Rev.* B **81** 161410

[106] Warner J H, Mukai M and Kirkland A I 2012 Atomic structure of ABC rhombohedral stacked trilayer graphene *ACS Nano* **6** 5680

[107] Biedermann L B, Bolen M L, Capano M A, Zemlyanov D and Reifenberger R G 2009 Insights into few-layer epitaxial graphene growth on 4H-SiC(000$\bar{1}$) substrates from STM studies *Phys. Rev.* B **79** 125411

[108] Borysiuk J, Soltys J and Piechota J 2011 Stacking sequence dependence of graphene layers on SiC (000$\bar{1}$) - experimental and theoretical investigation *J. Appl. Phys.* **109** 093523

[109] Reina A, Jia X, Ho J, Nezich D, Son H and Bulovic V *et al* 2009 Large area, few-layer graphene films on arbitrary substrates by chemical vapor deposition *Nano Lett.* **9** 30

[110] Liu L, Zhou H, Cheng R, Yu W J, Liu Y and Chen Y *et al* 2012 High-yield chemical vapor deposition growth of high-quality large-area AB-stacked bilayer graphene *ACS Nano.* **6** 8241

[111] Gomez T, Florez E, Rodriguez J A and Illas F 2011 Reactivity of transition metals (Pd, Pt, Cu, Ag, Au) toward molecular hydrogen dissociation: extended surfaces versus particles supported on TiC(001) or small is not always better and large is not always bad *J. Phys. Chem.* C **115** 11666

[112] Zhang X, Li H and Ding F 2014 Self-assembly of carbon atoms on transition metal surfaces-chemical vapor deposition growth mechanism of graphene *Adv. Mater.* **26** 5488

[113] Zhou Z, Habenicht B F and Guo Q 2013 Graphene moiré structure grown on a pseudomorphic metal overlayer supported on Ru(0001) *Surf. Sci.* **611** 67

[114] Kim K S, Zhao Y, Jang H, Lee S Y, Kim J M and Kim K S *et al* 2009 Large-scale pattern growth of graphene films for stretchable transparent electrodes *Nature* **457** 706

[115] Li X, Cai W, An J, Kim S, Nah J and Yang D *et al* 2009 Large-area synthesis of high-quality and uniform graphene films on copper foils *Science* **324** 1312

[116] Zhou H, Yu W J, Liu L, Cheng R, Chen Y and Huang X *et al* 2013 Chemical vapour deposition growth of large single crystals of monolayer and bilayer graphene *Nat. Commun.* **4** 2096

[117] Su C Y, Lu A Y, Wu C Y, Li Y T, Liu K K and Zhang W *et al* 2011 Direct formation of wafer scale graphene thin layers on insulating substrates by chemical vapor deposition *Nano Lett.* **11** 3612

[118] Bae S, Kim H, Lee Y, Xu X, Park J S and Zheng Y *et al* 2010 Roll-to-roll production of 30-inch graphene films for transparent electrodes *Nat. Nanotechnol.* **5** 574

[119] Vijayaraghavan R K, Gaman C, Jose B, Mccoy A P, Cafolla T and McNally P J *et al* 2016 Pulsed-plasma physical vapor deposition approach toward the facile synthesis of multilayer and monolayer graphene for anticoagulation applications *ACS Appl. Mater. Interfaces* **8** 4878

[120] Wu X Y, Zhong G F, D'Arsie L, Sugime H, Esconjauregui S and Robertson A W *et al* 2016 Growth of continuous monolayer graphene with millimeter-sized domains using industrially safe conditions *Sci. Rep.* **6** 21152

[121] Bosca A, Pedros J, Martinez J, Palacios T and Calle F 2016 Automatic graphene transfer system for improved material quality and efficiency *Sci. Rep.* **6** 21676

[122] Lee J K, Lee S C, Ahn J P, Kim S C, Wilson J I B and John P 2008 The growth of AAAA graphite on (111) diamond *J. Chem. Phys.* **129** 234709

[123] Que Y, Xiao W, Hhen C, Wang D, Du S and Gao H J 2015 Stacking-dependent electronic property of trilayer graphene epitaxially grown on Ru(0001) *Appl. Phys. Lett.* **107** 263101

[124] Ho Y H, Tsai S J, Lin M F and Su W P 2013 Unusual Landau levels in biased bilayer Bernal graphene *Phys. Rev.* B **87** 075417

[125] Lin Y P, Lin C Y, Chang C P and Lin M F 2015 Electric-field-induced rich magneto-absorption spectra of ABC-stacked trilayer graphene *RSC Adv.* **5** 80410

[126] Tran N T T, Lin S Y, Glukhova O E and Lin M F 2015 Configuration-induced rich electronic properties of bilayer graphene *J. Phys. Chem.* C **119** 10623

[127] Li T, Eugenio C, Daniele C, Carlo G, Marco F, Madan D, Alessandro M and Deji A 2015 Silicene field-effect transistors operating at room temperature *Nat. Nanotechnol.* **10** 227–31

[128] Vogt P, Padova P D, Quaresima C, Avila J, Frantzeskakis E, Asensio M C, Resta A, Ealet B and Lay G L 2012 Silicene: compelling experimental evidence for graphenelike two-dimensional silicon *Phys. Rev. Lett.* **108** 155501

[129] Aufray B, Kara A, Vizzini S, Oughaddou H, Landri C, Ealet B and Lay G L 2010 Graphene-like silicon nanoribbons on Ag(110): a possible formation of silicene *Appl. Phy. Lett.* **96** 183102

[130] Fleurence A, Friedlein R, Ozaki T, Kawai H, Wang Y and Yamada-Takamura Y 2012 Experimental evidence for epitaxial silicene on diboride thin films *Phys. Rev. Lett.* **108** 245501

[131] Meng L *et al* 2013 Buckled silicene formation on Ir(111) *Nano Lett.* **13** 685–90

[132] Li L F, Lu S Z, Pan J B, Qin Z H, Wang Y Q, Wang Y L, Cao G Y, Du S X and Gao H J 2014 Buckled germanene formation on Pt(111) *Adv. Mater.* **26** 4820–28

[133] Derivaz M, Dentel D, Stephan R, Hanf M C, Mehdaoui A, Sonnet P and Pirri C 2015 Continuous germanene layer on Al(111) *Nano Lett.* **15** 2510–16

[134] Dávila M E, Xian L, Cahangirov S, Rubio A and Lay G L 2014 Germanene: a novel two-dimensional germanium allotrope akin to graphene and silicene *New J. Phys.* **16** 095002

[135] Liu C C, Jiang H and Yao Y 2011 Low-energy effective Hamiltonian involving spin-orbit coupling in silicene and two-dimensional germanium and tin *Phys. Rev.* B **84** 195430

[136] Liu C C, Feng W and Yao Y 2011 Quantum spin Hall effect in silicene and two-dimensional germanium *Phys. Rev. Lett.* **107** 076802

[137] Ezawa M 2012 A topological insulator and helical zero mode in silicene under an inhomogeneous electric field *New J. Phys.* **14** 033003

[138] Drummond N D, Zólyomi V and Fal'ko V I 2012 Electrically tunable band gap in silicene *Phys. Rev.* B **85** 075423

[139] Zhu F, Chen W, Xu Y, Gao C, Guan D, Liu C, Qian D, Zhang S C and Jia J 2015 Epitaxial growth of two-dimensional stanene *Nat. Mater.* **14** 1020–25

[140] Xu Y, Yan B, Zhang H J, Wang J, Xu G, Tang P, Duan W and Zhang S C 2013 Large-gap quantum spin Hall insulators in tin films *Phys. Rev. Lett.* **111** 136804

[141] Liu H, Du Y, Denga Y and Ye P D 2015 Semiconducting black phosphorus: synthesis, transport properties and electronic applications *Chem. Soc. Rev.* **44** 2732

[142] Li L, Yu Y, Ye G J, Ge Q, Ou X, Wu H, Feng D, Chen X H and Zhang Y 2014 Black phosphorus field-effect transistors *Nat. Nanotechnol.* **9** 372

[143] Liu H, Neal A T, Zhu Z, Luo Z, Xu X, Tománek D and Ye P D 2014 Phosphorene: an unexplored 2D semiconductor with a hole hole mobility *ACS Nano* **8** 4033

[144] Brent J R, Savjani N, Lewis E A, Haigh S J, Lewis D J and O'Brien P 2014 Production of few-layer phosphorene by liquid exfoliation of black phosphorus *Chem. Commun.* **50** 13338

[145] Yasaei P, Kumar B, Foroozan T, Wang C, Asadi M, Tuschel D, Indacochea J E, Klie R F and Salehi-Khojin A 2015 High-quality black phosphorus atomic layers by liquid-phase exfoliation *Adv. Mater.* **27** 1887

[146] Kang J, Wood J D, Wells S A, Lee J-H, Liu X, Chen K S and Hersam M C 2015 Solvent exfoliation of electronic-grade, two-dimensional black phosphorus *ACS Nano* **9** 3596

[147] Lange S, Schmidt P and Nilges T 2007 Au_3SnP_7@black phosphorus: an easy access to black phosphorus *Inorg. Chem.* **46** 4028

[148] Nilges T, Kersting M and Pfeifer T 2008 A fast low-pressure transport route to large black phosphorus single crystals *J. Solid State Chem.* **181** 1707

[149] Kopf M, Eckstein N, Pfister D, Grotz C, Kruger I, Greiwe M, Hansen T, Kohlmann H and Nilges T 2014 Access and *in situ* growth of phosphorene-precursor black phosphorus *J. Cryst. Growth* **405** 6

[150] Rudenko A N and Katsnelson M I 2014 Quasiparticle band structure and tight-binding model for single- and bilayer black phosphorus *Phys. Rev.* B **89** 201408

[151] Low T, Rodin A S, Carvalho A, Jiang Y, Wang H, Xia F and Castro Neto A H 2014 Tunable optical properties of multilayer black phosphorus thin films *Phys. Rev.* B **90** 075434

[152] Dolui K and Quek S Y 2015 Quantum-confinement and structural anisotropy result in electrically-tunable Dirac cone in few-layer black phosphorus *Sci. Rep.* **5** 11699

[153] Liu Q, Zhang X, Abdalla L B, Fazzio A and Zunger A 2015 Switching a normal insulator into a topological insulator via electric field with application to phosphorene *Nano Lett.* **15** 1222

[154] Novoselov K S, Jiang D, Schedin F, Booth T J, Khotkevich V V, Morozov S V and Geim A K 2005 Two-dimensional atomic crystals *Proc. Natl. Acad. Sci. U. S. A.* **102** 10451

[155] Yin Z Y, Li H, Jiang L, Shi Y M, Sun Y H, Lu G, Zhang Q, Chen X D and Zhang H 2012 Single-layer MoS_2 phototransistors *ACS Nano* **6** 74

[156] Li H, Wu J, Yin Z and Zhang H 2014 Preparation and applicatons of mechanically exfoliated single layer and multilayer MoS_2 and WSe_2 nanosheets *Acc. Chem. Res.* **47** 1067–75

[157] Mak K F, Lee C, Hone J, Shan J and Heinz T F 2010 Atomically thin MoS_2: a new direct-gap semiconductor *Phys. Rev. Lett.* **105** 136805

[158] Coleman J N, Lotya M, O'Neill A, Bergin S D, King P J and Khan U *et al* 2011 Two-dimensional nanosheets produced by liquid exfoliation of layered materials *Science* **331** 568–71

[159] Zhou K G, Mao N N, Wang H X, Peng Y and Zhang H L 2011 A mixed-solvent strategy for efficient exfoliation of inorganic graphene analogues *Angew. Chem. Int. Ed* **50** 10839

[160] Lee Y H *et al* 2012 Synthesis of large-area MoS_2 atomic layers with chemical vapor deposition *Adv. Mater.* **24** 2320–5

[161] Najmaei S, Liu Z, Zhou W, Zou X, Shi G, Lei S, Yakobson B I, Idrobo J C, Ajayan P M and Lou J 2013 Vapour phase growth and grain boundary structure of molybdenum disulphide atomic layers *Nat. Mater.* **12** 754–9

[162] Liu B, Fathi M, Chen L, Abbas A, Ma Y and Zhou C 2015 Chemical vapor deposition growth of monolayer WSe_2 with tunable device characteristics and growth mechanism study *ACS Nano* **9** 6119–27

[163] Shaw J C, Zhou H, Chen Y, Weiss N O, Liu Y, Huang Y and Duan X 2014 Chemical vapor deposition growth of monolayer $MoSe_2$ nanosheets *Nano Res.* **7** 511–7

[164] Liu K K *et al* 2012 Growth of large-area and highly crystalline MoS_2 thin layers on insulating substrates *Nano Lett.* **12** 1538–44

[165] Yoon Y, Ganapathi K and Salahuddin S 2011 How good can monolayer MoS_2 transistors be? *Nano Lett.* **11** 3768–73

[166] Wang H, Yu L, Lee Y H, Shi Y, Hsu A, Chin M L, Li L J, Dubey M, Kong J and Palacios T 2012 Integrated circuits based on bilayer MoS_2 transistors *Nano Lett.* **12** 4674–80

[167] Zhang Y J, Ye J T, Matsuhashi Y and Iwasa Y 2012 Ambipolar MoS_2 thin flake transistors *Nano Lett.* **12** 1136

[168] Wang Q H, Kalantar-Zadeh K, Kis A, Coleman J N and Strano M S 2012 Electronics and optoelectronics of two-dimensional transition metal dichalcogenides *Nat. Nanotechnol.* **7** 699–712

[169] Xiao D, Liu G B, Feng W, Xu X and Yao W 2012 Coupled spin and valley physics in monolayers of MoS_2 and other group-VI dichalcogenides *Phys. Rev. Lett.* **108** 196802

[170] Yao W, Xiao D and Niu Q 2008 Valley-dependent optoelectronics from inversion symmetry breaking *Phys. Rev. B* **77** 235406

[171] Splendiani A, Sun L, Zhang Y, Li T, Kim J, Chim C Y, Galli G and Wang F 2010 Emerging photoluminescence in monolayer MoS_2 *Nano Lett.* **10** 1271

[172] Ross J S *et al* 2013 Electrical control of neutral and charged excitons in a monolayer semiconductor *Nat. Commun.* **4** 1474

[173] Cao T *et al* 2012 Valley-selective circular dichroism of monolayer molybdenum disulphide *Nat. Commun.* **3** 887

[174] Liu G B, Shan W-Y, Yao Y, Yao W and Xiao D 2013 Three-band tight-binding model for monolayers of group-VIB transition metal dichalcogenides *Phys. Rev.* B **88** 085433

[175] Luican A, Li G H, Reina A, Kong J, Nair R R, Novoselov K S, Geim A K and Andrei E Y 2011 Single-layer behaviour and its breakdown in twisted graphene layers *Phys. Rev. Lett.* **106** 126802

[176] Li G, Luican A, Dos Santos J L, Neto A C, Reina A, Kong J and Andrei E Y 2010 Observation of Van Hove singularities in twisted graphene layers *Nat. Phys.* **6** 109

[177] Cherkez V, de Laissardiere G T, Mallet P and Veuillen J-Y 2015 Van Hove singularities in doped twisted graphene bilayers studied by scanning tunneling spectroscopy *Phys. Rev.* B **91** 155428

[178] Lauffer P, Emtsev K, Graupner R, Seyller T, Ley L and Reshanov S *et al* 2008 Atomic and electronic structure of few-layer graphene on SiC (0001) studied with scanning tunneling microscopy and spectroscopy *Phys. Rev.* B **77** 155426

[179] Yankowitz M, Wang F, Lau C N and LeRoy B J 2013 Local spectroscopy of the electrically tunable band gap in trilayer graphene *Phys. Rev.* B **87** 165102

[180] Pierucci D, Sediri H, Hajlaoui M, Girard J-C, Brumme T and Calandra M *et al* 2015 Evidence for flat bands near the Fermi level in epitaxial rhombohedral multilayer graphene *ACS Nano* **9** 5432

[181] Li G H, Luican A and Andrei E Y 2009 Scanning tunneling spectroscopy of graphene on graphite *Phys. Rev. Lett.* **102** 176804

[182] Klusek Z 1999 Investigations of splitting of the bands in graphite by scanning tunneling spectroscopy *Appl. Surf. Sci.* **151** 251

[183] Huang H, Wei D, Sun J, Wong S L, Feng Y P and Neto A H C *et al* 2012 Spatially resolved electronic structures of atomically precise armchair graphene nanoribbons *Sci. Rep.* **2** 231

[184] Söde H, Talirz L, Gröning O, Pignedoli C A, Berger R and Feng X *et al* 2015 Electronic band dispersion of graphene nanoribbons via Fourier transformed scanning tunneling spectroscopy *Phys. Rev.* B **91** 045429

[185] Chen Y C, De Oteyza D G, Pedramrazi Z, Chen C, Fischer F R and Crommie M F 2013 Tuning the band gap of graphene nanoribbons synthesized from molecular precursors *ACS Nano* **7** 6123

[186] Wilder J W G, Venema L C, Rinzler A G, Smalley R E and Dekker C 1998 Electronic structure of atomically resolved carbon nanotubes *Nature* **391** 59

[187] Odom T W, Huang J-L, Kim P and Lieber C M 1998 Atomic structure and electronic properties of single-walled carbon nanotubes *Nature* **391** 62

[188] Yin L J, Li S Y, Qiao J B, Nie J C and He L 2015 Landau quantization in graphene monolayer, Bernal bilayer, and Bernal trilayer on graphite surface *Phys. Rev.* B **91** 115405

[189] Rutter G M, Jung S, Klimov N N, Newell D B, Zhitenev N B and Stroscio J A 2011 Microscopic polarization in bilayer graphene *Nat. Phys.* **7** 649

[190] Yin L J, Zhang Y, Qiao J B, Li S Y and He L 2016 Experimental observation of surface states and Landau levels bending in bilayer graphene *Phys. Rev.* B **93** 125422

[191] Matsui T, Kambara H, Niimi Y, Tagami K, Tsukada M and Fukuyama Hiroshi 2005 STS observations of Landau levels at graphite surfaces *Phys. Rev. Lett.* **94** 226403

[192] Li G H and Andrei E Y 2007 Observation of Landau levels of Dirac fermions in graphite *Nat. Phys.* **3** 623

[193] Orlita M, Faugeras C, Plochocka P, Neugebauer P, Martinez G and Maude D K *et al* 2008 Approaching the Dirac point in high-mobility multilayer expitaxial graphene *Phys. Rev. Lett.* **101** 267601

[194] Orlita M, Faugeras C, Schneider J M, Martinez G, Maude D K and Potemski M 2009 Graphite from the viewpoint of Landau level spectroscopy: an effective graphene bilayer and monolayer *Phys. Rev. Lett.* **102** 166401

[195] Orlita M, Faugeras C, Martinez G, Maude D K, Sadowski M L and Potemski M 2008 Dirac fermions at the H point of graphite: magnetotransmission studies *Phys. Rev. Lett.* **100** 136403

[196] Berciaud S, Potemski M and Faugeras C 2014 Probing electronic excitations in mono- to pentalayer graphene by micro magneto-Raman spectroscopy *Nano Lett.* **14** 4548

[197] Carmona H A, Geim A K, Nogaret A, Main P C, Foster T J, Henini M, Beaumont S P and Blamire M G 1995 Two dimensional electrons in a lateral magnetic superlattice *Phys. Rev. Lett.* **74** 3009

[198] Kato M, Endo A, Katsumoto S and Iye Y 1998 Two-dimensional electron gas under a spatially modulated magnetic field: a test ground for electron-electron scattering in a controlled environment *Phys. Rev.* B **58** 4876

[199] Kato M, Endo A, Sakairi M, Katsumoto S and Iye Y 1999 Electron-electron Umklapp process in two-dimensional electron gas under a spatially alternating magnetic field *J. Phys. Soc. Japan* **68** 1492

[200] Gerhardts R R, Weiss D and Klitzing K v 1989 Novel magnetoresistance oscillations in a periodically modulated two-dimensional electron gas *Phys. Rev. Lett.* **62** 1173

[201] Messica A, Soibel A, Meirav U, Stern A, Shtrikman H, Umansky V and Mahalu D 1997 Suppression of conductance in surface superlattices by temperature and electric field *Phys. Rev. Lett.* **78** 705

[202] Chiu Y H, Lai Y H, Ho J H, Chuu D S and Lin M F 2008 Electronic structure of a two-dimensional graphene monolayer in a spatially modulated magnetic field: Peierls tight-binding model *Phys. Rev.* B **77** 045407

[203] Ho J H, Lai Y H, Chiu Y H and Lin M F 2008 Modulation effects on Landau levels in a monolayer graphene *Nanotechnology* **19** 035712

[204] Chiu Y H, Ho J H, Chang C P, Chuu D S and Lin M F 2008 Low-frequency magneto-optical excitations of a graphene monolayer: Peierls tight-binding model and gradient approximation calculation *Phys. Rev.* B **78** 245411

[205] Park S and Sim H S 2008 Magnetic edge states in graphene in nonuniform magnetic fields *Phys. Rev.* B **77** 075433

IOP Publishing

Theory of Magnetoelectric Properties of 2D Systems

S C Chen, J Y Wu, C Y Lin and M F Lin

Chapter 2

The generalized tight-binding model

To fully understand the electronic properties of emergent layered materials, we propose the generalized tight-binding model to solve the various Hamiltonians under the magnetic and electric fields. The typical systems, graphene, silicene, germanene, tinene, phosphorene and MoS_2, are suitable for a model study. The planar/buckled/puckered and layered structures, with the distinct lattice symmetries, layer numbers and stacking configurations, are taken into consideration. Furthermore, the geometry- and atom-dominated interactions, the site energies, the single- or multi-orbital hopping integrals, the orbital-dependent SOCs, and the intralayer and interlayer atomic interactions, are included in the Hamiltonian. The field-created independent Hamiltonian matrix elements are derived in the analytic form, especially for the characteristics of this model (reliability, suitability, extension and combination), which are directly reflected in the delicate calculations of magneto-electronic properties, are discussed in detail. How to use it in studying other essential properties is worthy of a closer discussion, such as the combinations with the modified random-phase approximation and the static/dynamic Kubo formula. Finally, the diagonalization of a huge Hermitian matrix can be efficiently done by using the band-like one and the features of the LL spatial distributions.

2.1 Monolayer graphene

Monolayer graphene, as shown in figure 1.1(a), has a honeycomb lattice with the C–C bond length $b = 1.42$ Å. In the absence of external fields, there are two carbon atoms, the A and B atoms, in a primitive unit cell. The low- and middle-energy essential properties are dominated by the $2p_z$ orbitals. This indicates that the Bloch wave function is a linear superposition of two tight-binding functions due to the

doi:10.1088/978-0-7503-1674-3ch2
© IOP Publishing Ltd 2017

periodical orbitals. $|\Psi_{\mathbf{k}}| = c_A|A_{\mathbf{k}}\rangle + c_B|B_{\mathbf{k}}\rangle$, where $|A_{\mathbf{k}}\rangle$ and $|B_{\mathbf{k}}\rangle$, respectively, stand for the tight-binding functions of the A and B atoms:

$$|A_{\mathbf{k}}\rangle = \frac{1}{\sqrt{N_A}} \sum_{J=1}^{N_A} \exp(i\mathbf{k} \cdot \mathbf{R}_{A_J})\chi(\mathbf{r} - \mathbf{R}_{A_J});$$

$$|B_{\mathbf{k}}\rangle = \frac{1}{\sqrt{N_B}} \sum_{J=1}^{N_B} \exp(i\mathbf{k} \cdot \mathbf{R}_{B_J})\chi(\mathbf{r} - \mathbf{R}_{B_J}).$$

$$(2.1)$$

$N_A(N_B)$ is the total number of A(B) atoms, and $\chi(\mathbf{r} - \mathbf{R}_{A_J})$ $(\chi(\mathbf{r} - \mathbf{R}_{B_J}))$ is the normalized $2p_z$ orbital wave function centered at $\mathbf{R}_{A_J}(\mathbf{R}_{B_J})$. When the nearest-neighbor atomic interaction (hopping integral $\gamma_0 = -2.6$ eV [1]) is taken into consideration, it is sufficient in understanding the main features of the π-electronic structure. As a result, the Hamiltonian matrix elements built from the space spanned by the two tight-binding functions are expressed as

$$\langle A_{\mathbf{k}}|H|A_{\mathbf{k}}\rangle = \langle B_{\mathbf{k}}|H|B_{\mathbf{k}}\rangle = 0,$$
$$\langle A_{\mathbf{k}}|H|B_{\mathbf{k}}\rangle = \gamma_0(\exp(-ik_xb) + 2\exp(ik_xb/2)\cos(k_y\sqrt{3}\,b/2));$$
$$\langle B_{\mathbf{k}}|H|A_{\mathbf{k}}\rangle = \langle A_{\mathbf{k}}|H|B_{\mathbf{k}}\rangle^*.$$

$$(2.2)$$

In the presence of a uniform perpendicular magnetic field $\mathbf{B} = \mathbf{B}_z\hat{z}$, the vector potential, being chosen as $\mathbf{A}(r) = \mathbf{B}_z x\hat{y}$, can create a position-dependent phase of $G_n = \frac{2\pi}{\Phi_0} \int_{\mathbf{R}_n}^{\mathbf{r}} \mathbf{A} \cdot d\mathbf{l}$ in the tight-binding functions. $\Phi_0 = hc/e$ (4.1356×10^{-15}[T $\cdot$ m^2]) is the flux quantum. The nearest-neighbor hopping integral is modulated as

$$\langle A_{J,\mathbf{k}}|H|B_{I,\mathbf{k}}\rangle = \gamma_0 \exp\left\{i\left[\mathbf{k} \cdot (\mathbf{R}_I - \mathbf{R}_J) + \frac{2\pi}{\Phi_0} \int_{\mathbf{R}_{B_I}}^{\mathbf{R}_{A_J}} \mathbf{A} \cdot d\mathbf{l}\right]\right\}.$$

$$(2.3)$$

Owing to the periodicity of the Peierls phase $\frac{2\pi}{\Phi_0} \int_{\mathbf{R}_{B_I}}^{\mathbf{R}_{A_J}} \mathbf{A} \cdot d\mathbf{l}$, a hexagonal primitive unit cell becomes an enlarged rectangle along the x-direction (the armchair direction), as clearly indicated in figure 1.1(a). Furthermore, the hexagonal first Brillouin zone (figure 1.1(b)) is changed into a very small rectangle (figure 1.1(c); an area of $4\pi^2/3\sqrt{3}\,b^2R_B$). R_B, which is related to the period along $\hat{x}$, is defined as the ratio of Φ_0 versus magnetic flux through each hexagon (Φ), e.g. $R_B = 2 \times 10^3$ at $B_z = 40$ T. Accordingly, the B_z-enlarged rectangular cell includes $4R_B$ atoms ($2R_B$A and $2R_B$B atoms); its length along the x-direction is $l_x = 3R_Bb$. This implies that the Bloch wave functions under a uniform magnetic field can be expressed by the linear superposition of the $4R_B$ tight-binding functions in the rectangular unit cell: $|A_{1\mathbf{k}}\rangle$, $|B_{1\mathbf{k}}\rangle$, $|A_{2\mathbf{k}}\rangle$, $|B_{2\mathbf{k}}\rangle$, $\ldots|A_{2R_B-1\mathbf{k}}\rangle$, $|B_{2R_B-1\mathbf{k}}\rangle$, $|A_{2R_B\mathbf{k}}\rangle$; $|B_{2R_B\mathbf{k}}\rangle$. Under these bases, the nonzero and independent Hamiltonian matrix elements are given by

$$\langle A_{J,\mathbf{k}}|H|B_{I,\mathbf{k}}\rangle = \gamma_0(t_{1,J} + t_{2,J})\delta_{J,I} + \gamma_0 t_3\delta_{J,I+1},$$

$$(2.4)$$

where the vector-potential-induced phase terms in the hopping integrals are

$$t_{1,J} = \exp\left\{i\left[(k_x b/2 + k_y \sqrt{3}\,b/2) - \pi\frac{\Phi}{\Phi_0}\left(J - 1 + \frac{1}{6}\right)\right]\right\},$$

$$t_{2,J} = \exp\left\{i\left[(k_x b/2 - k_y \sqrt{3}\,b/2) + \pi\frac{\Phi}{\Phi_0}\left(J - 1 + \frac{1}{6}\right)\right]\right\}; \tag{2.5}$$

$$t_3 = \exp(ik_y b).$$

To solve the $4R_B \times 4R_B$ Hamiltonian matrix more efficiently, a band-like Hamiltonian matrix is introduced by rearranging the bases as the following sequence: $|A_{1\mathbf{k}}\rangle$, $|B_{2R_B\mathbf{k}}\rangle$, $|B_{1\mathbf{k}}\rangle$, $|A_{2R_B\mathbf{k}}\rangle$, $|A_{2\mathbf{k}}\rangle$, $|B_{2R_B-1\mathbf{k}}\rangle$, $|B_{2\mathbf{k}}\rangle$, $|A_{2R_B-1\mathbf{k}}\rangle$.... As to the $(k_x = 0, k_y = 0)$ magnetic states, the Hamiltonian matrix elements are real numbers. This numerical characteristic is also revealed in layered graphenes with normal stacking configurations. State energy and wave function, $E^{c,v}$ and $\Psi^{c,v}$, are obtained from diagonalizing the Hamiltonian matrix, where c and v, respectively, represent conduction and valence states. In addition, similar equations, as expressed in equations (2.3)–(2.5), can be derived for a modulated/composite magnetic field.

According to the exact diagonalization scheme, the Landau wavefunction can be expressed as

$$|\Psi_{\mathbf{k}}^{c,v}\rangle = \sum_{J=1}^{2R_B} A_J|A_{J\mathbf{k}}\rangle + B_J|B_{J\mathbf{k}}\rangle, \tag{2.6}$$

where the subenvelope function $A_J(B_J)$ corresponds to the probability amplitude of the tight-binding function based on the Jth $A(B)$ atom in the enlarged cell. The spatial distribution of subenvelope function will present a special oscillation mode, being very useful in the characterization of each LL. This method has well defined the quantum number of LLs in layered graphenes and other systems discussed in this work. The generalization of the magnetic wave function is valuable in understanding other physical properties, such as the critical mechanisms of quantum transports [2], magneto-optical spectra [3–9] and Coulomb excitations [10–19].

2.2 Tetra-layer graphene

ABC-stacked graphene has an interlayer distance of $d = 3.37$ Å, as sketched in figure 2.1(a). For an N-layer system, $2N$ carbon atoms are included in a primitive unit cell, in which two sublattice atoms in the lth layer are denoted as A^l and B^l. Each graphene sheet is shifted by a distance of b along the armchair direction with respect to the adjacent layer. The sublattice A(B) of one layer is situated directly above the A atom of the adjacent lower layer, while the sublattice B(A) lies above the center of its hexagon. The layer-dependent Hamiltonian is characterized by the intralayer and the interlayer atomic interactions β_i's (figure 2.1(a)), where β_0 represents the nearest-neighbor hopping integral within the same layer; $(\beta_1, \beta_3, \beta_4)$ are between adjacent layers; β_2 and β_5 are related to the next-neighboring layers. β_1 and β_2 are couplings between two vertical sites, and $(\beta_3, \beta_4, \beta_5)$ belong to

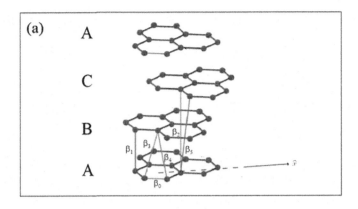

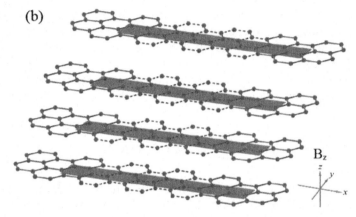

Figure 2.1. (a) Geometric structure and (b) B_z-dependent unit cell of the ABC-stacked tetralayer graphene. β_i's denote the intralayer and interlayer hopping integrals.

non-vertical interactions. The values of hopping integrals are as follows: $\beta_0 = -3.16$ eV, $\beta_1 = 0.36$ eV, $\beta_2 = -0.01$ eV, $\beta_3 = 0.32$ eV, $\beta_4 = 0.03$ eV, and $\beta_5 = 0.0065$ eV [20].

As to tetra-layer ABC stacking, the zero-field Hamiltonian, being associated with eight carbon atoms in a unit cell, has the block matrix form

$$H_{ABC} = \begin{pmatrix} H_1 & H_2 & H_3 & 0 \\ H_2^* & H_1 & H_2 & H_3 \\ H_3^* & H_2^* & H_1 & H_2 \\ 0 & H_3^* & H_2^* & H_1 \end{pmatrix},$$

(2.7)

where three independent 2×2 matrices are

$$H_1 = \begin{pmatrix} 0 & \beta_0 f(k_x, k_y) \\ \beta_0 f^*(k_x, k_y) & 0 \end{pmatrix},$$

(2.8)

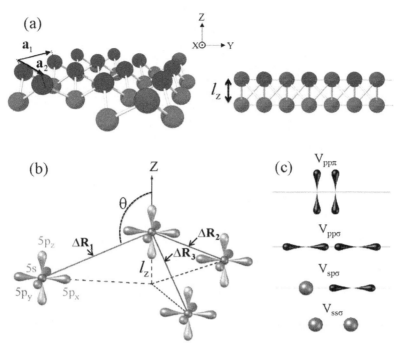

Figure 2.2. (a) Geometric structure of the buckled (silicene, germanene, tinene) with (b) the sp^3 bondings, or (c) significant orbital hybridizations.

$$H_2 = \begin{pmatrix} \beta_4 f^*(k_x, k_y) & \beta_1 \\ \beta_3 f(k_x, k_y) & \beta_4 f^*(k_x, k_y) \end{pmatrix}; \tag{2.9}$$

$$H_3 = \begin{pmatrix} \beta_5 f(k_x, k_y) & \beta_5 f^*(k_x, k_y) \\ \beta_2 & \beta_5 f(k_x, k_y) \end{pmatrix}. \tag{2.10}$$

The summation of the nearest-neighbor phase terms is $f(k_x, k_y) = \sum_{j=1}^{3}$ $\exp(i\mathbf{k} \cdot \mathbf{r}_j) = \exp(ibk_x) + \exp(ibk_x/2)\cos(\sqrt{3}\,bk_y/2)$. The diagonal matrix elements are changed from zero into finite values in the presence of a uniform perpendicular electric field; that is, E_z leads to the Coulomb potential energy differences in distinct layers. For experimental measurements, this field can be created by the application of gate voltage (V_z) across the layered systems.

The periodical Peierls phase in tetra-layer ABC stacking can create an enlarged rectangle with $16R_B$ atoms. The magnetic Hamiltonian is constructed from the $16R_B$ tight-binding functions $\{|A_{1\mathbf{k}}^1\rangle, |B_{1\mathbf{k}}^1\rangle, |A_{1\mathbf{k}}^2\rangle, |B_{1\mathbf{k}}^2\rangle \dots |A_{1\mathbf{k}}^4\rangle, |B_{1\mathbf{k}}^4\rangle \dots |A_{2R_B\mathbf{k}}^4\rangle, |B_{2R_B\mathbf{k}}^4\rangle\}$ based on the periodical atoms in the rectangular unit cell (figure 2.1(b)). By the detailed calculations, the non-vanishing Hamiltonian matrix elements associated with the hopping integrals β_i's cover

$$\left\langle B_{J\mathbf{k}}^1 \middle| H \middle| A_{I\mathbf{k}}^1 \right\rangle = \left\langle B_{J\mathbf{k}}^4 \middle| H \middle| A_{I\mathbf{k}}^4 \right\rangle = \beta_0(t_{1,I}\delta_{I,J} + q\delta_{I,J+1}), \tag{2.11}$$

$$\left\langle B_{J\mathbf{k}}^2 \middle| H \middle| A_{I\mathbf{k}}^2 \right\rangle = \beta_0(t_{3,I}\delta_{I,J-1} + q\delta_{I,J}), \tag{2.12}$$

$$\left\langle B_{J\mathbf{k}}^3 \middle| H \middle| A_{I\mathbf{k}}^3 \right\rangle = \beta_0(t_{2,I}\delta_{I,J} + q\delta_{I,J+1}), \tag{2.13}$$

$$\left\langle A_{J\mathbf{k}}^1 \middle| H \middle| B_{I\mathbf{k}}^2 \right\rangle = \left\langle A_{J\mathbf{k}}^2 \middle| H \middle| B_{I\mathbf{k}}^3 \right\rangle = \left\langle A_{J\mathbf{k}}^3 \middle| H \middle| B_{I\mathbf{k}}^4 \right\rangle = \beta_1\delta_{I,J}, \tag{2.14}$$

$$\left\langle B_{J\mathbf{k}}^1 \middle| H \middle| A_{I\mathbf{k}}^3 \right\rangle = \left\langle B_{J\mathbf{k}}^2 \middle| H \middle| A_{I\mathbf{k}}^4 \right\rangle = \beta_2\delta_{I,J}, \tag{2.15}$$

$$\left\langle A_{J\mathbf{k}}^2 \middle| H \middle| B_{I\mathbf{k}}^1 \right\rangle = \beta_3(t_{2,I}\delta_{I,J} + q\delta_{I,J+1}), \tag{2.16}$$

$$\left\langle A_{J\mathbf{k}}^4 \middle| H \middle| B_{I\mathbf{k}}^3 \right\rangle = \beta_3(t_{3,I}\delta_{I,J-1} + q\delta_{I,J}), \tag{2.17}$$

$$\left\langle A_{J\mathbf{k}}^3 \middle| H \middle| B_{I\mathbf{k}}^2 \right\rangle = \beta_3(t_{1,I}\delta_{I,J} + q\delta_{I,J+1}), \tag{2.18}$$

$$\left\langle B_{j\mathbf{k}}^1 \middle| H \middle| B_{I\mathbf{k}}^2 \right\rangle = \left\langle A_{j\mathbf{k}}^3 \middle| H \middle| A_{I\mathbf{k}}^4 \right\rangle = \beta_4(t_{1,I}\delta_{I,J} + q\delta_{I,J+1}), \tag{2.19}$$

$$\left\langle A_{j\mathbf{k}}^1 \middle| H \middle| A_{I\mathbf{k}}^2 \right\rangle = \left\langle B_{j\mathbf{k}}^2 \middle| H \middle| B_{i\mathbf{k}}^3 \right\rangle = \beta_4(t_{3,I}\delta_{I,J-1} + q\delta_{I,J}), \tag{2.20}$$

$$\left\langle A_{J\mathbf{k}}^2 \middle| H \middle| A_{I\mathbf{k}}^3 \right\rangle = \left\langle B_{J\mathbf{k}}^3 \middle| H \middle| B_{I\mathbf{k}}^4 \right\rangle = \beta_4(t_{2,I}\delta_{I,J} + q\delta_{I,J+1}), \tag{2.21}$$

$$\left\langle B_{J\mathbf{k}}^3 \middle| H \middle| B_{I\mathbf{k}}^1 \right\rangle = \left\langle A_{J\mathbf{k}}^2 \middle| H \middle| B_{I\mathbf{k}}^4 \right\rangle = \beta_5(t_{2,I}\delta_{I,J} + q\delta_{I,J+1}), \tag{2.22}$$

$$\left\langle A_{J\mathbf{k}}^3 \middle| H \middle| A_{I\mathbf{k}}^1 \right\rangle = \left\langle B_{J\mathbf{k}}^4 \middle| H \middle| B_{I\mathbf{k}}^2 \right\rangle = \beta_5(t_{1,I}\delta_{I,J} + q\delta_{I,J+1}), \tag{2.23}$$

$$\left\langle A_{J\mathbf{k}}^1 \middle| H \middle| B_{I\mathbf{k}}^3 \right\rangle = \left\langle A_{J\mathbf{k}}^4 \middle| H \middle| A_{I\mathbf{k}}^2 \right\rangle = \beta_5(t_{3,I}\delta_{I,J-1} + q\delta_{I,J}). \tag{2.24}$$

The four independent phase terms are

$$t_{1,I} = \exp\left\{i\left[-(k_x b/2) - \left(\sqrt{3}\,k_y b/2\right) + \pi\Phi(I - 1 + 1/6)\right]\right\}$$
$$+ \exp\left\{i\left[-(k_x b/2) + \left(\sqrt{3}\,k_y b/2\right) - \pi\Phi(I - 1 + 1/6)\right]\right\},$$

$$t_{2,I} = \exp\left\{i\left[-(k_x b/2) - \left(\sqrt{3}\,k_y b/2\right) + \pi\Phi(I - 1 + 3/6)\right]\right\}$$
$$+ \exp\left\{i\left[-(k_x b/2) + \left(\sqrt{3}\,k_y b/2\right) - \pi\Phi(I - 1 + 3/6)\right]\right\},$$

$$t_{3,I} = \exp\left\{i\left[-(k_x b/2) - \left(\sqrt{3}\,k_y b/2\right) + \pi\Phi(I - 1 + 5/6)\right]\right\}$$
$$+ \exp\left\{i\left[-(k_x b/2) + \left(\sqrt{3}\,k_y b/2\right) - \pi\Phi(I - 1 + 5/6)\right]\right\},$$

and $q = \exp\{ik_x b\}$.

In order to enhance the computation efficiency, we choose the bases of $\{|A_{1k}^1\rangle$, $|B_{1k}^2\rangle$, $|A_{1k}^3\rangle$, $|B_{1k}^4\rangle$, $|B_{1k}^1\rangle$, $|A_{1k}^2\rangle$, $|B_{1k}^3\rangle$, $|A_{1k}^4\rangle$, ... $|B_{2R_Bk}^1\rangle$, $|A_{2R_Bk}^2\rangle$, $|B_{2R_Bk}^3\rangle$, $|A_{2R_Bk}^4\rangle\}$ to arrange the Hamiltonian as a band-like symmetric matrix. This allows an efficient numerical solution of the eigenvalues and eigenfunctions, even for a small magnetic field strength and a huge R_B. The analytic formulas for the Hamiltonian matrix elements could also be obtained for any N-layer systems with the normal stacking configurations (AAA, ABA, ABC and AAB). Furthermore, such matrix elements remain similar under a composite B_z- and E_z-field.

2.3 Tinene

The generalized tight-binding model is further developed to include sp^3 orbital bonding, SOC, and the external fields simultaneously. Monolayer buckled tinene, as shown in figure 2.2(a), consists of two equivalent A and B sublattices, respectively, situated at two parallel planes with a separation of $l_z = 0.417$ Å. The primitive unit vectors are $\mathbf{a_1}$ and $\mathbf{a_2}$ with a lattice constant of a = 4.7 Å, and the angle between the Sn-Sn bond and the z-axis is $\theta = 107.1°$ (figure 2.2(b)). The 5s orbital energy is $E_{5s} = -6.23$ eV below that of the 5p orbitals taken as zero ($E_{5p} = 0$). The Slater–Koster hopping parameters in the sp^3 bonding are $V_{ss\sigma} = -2.62$ eV, $V_{sp\sigma} = 2.65$ eV, $V_{pp\sigma} = 1.49$ eV, and $V_{pp\pi} = -0.79$ eV. Such significant chemical bondings are clearly indicated in figure 2.2(c). The SOC strength ($\lambda_{SOC} = 0.8$ eV) of tinene is predicted to be two orders of magnitude greater than that of graphene [21, 22]. In the orbital- and spin-dependent bases of $\left\{|5p_z^A\rangle, |5p_x^A\rangle, |5p_y^A\rangle, |5s^A\rangle, |5p_z^B\rangle, |5p_x^B\rangle, |5p_y^B\rangle, |5s^B\rangle\right\} \otimes \{\uparrow, \downarrow\}$, the Hamiltonian, with the nearest-neighbor atomic interactions, presents two types of site-dominated matrix elements:

$$\langle A_{om}|H|B_{o'm'}\rangle = \sum_{\Delta\mathbf{R}_J} \gamma_{oo'}^{\Delta\mathbf{R}_J} \exp i\mathbf{k}\cdot\Delta\mathbf{R}_J\delta_{mm'},$$

$$\langle A_{om}|H|A_{om}\rangle = \langle B_{om}|H|B_{om}\rangle = E_o,$$

(2.25)

where $o(o')$, $m(m')$, $\Delta\mathbf{R}_J$ and E_o stand for the atomic orbital, spin, the translation vector of the nearest-neighbor atom, and site energy, respectively. The nearest-neighbor hopping integral ($\gamma_{oo'}^{\Delta\mathbf{R}_J}$) depends on the type of atomic orbitals, $\Delta\mathbf{R}_J$, and θ. $\gamma_{oo'}^{\Delta\mathbf{R}_J}s$ for the three nearest-neighbor atoms and different atomic orbitals are given by

$$\gamma_{xx}^{\Delta R_1} = V_{pp\sigma} + (V_{pp\pi} - V_{pp\sigma})\cos^2\theta,$$

$$\gamma_{xx}^{\Delta R_2} = \gamma_{xx}^{\Delta R_3} = V_{pp\pi} + \frac{1}{4}(V_{pp\pi} - V_{pp\sigma})\sin^2\theta,$$

$$\gamma_{xy}^{\Delta R_1} = 0,$$

$$\gamma_{xy}^{\Delta R_2} = -\gamma_{xy}^{\Delta R_3} = \frac{\sqrt{3}}{4}(V_{pp\pi} - V_{pp\sigma})\sin^2\theta,$$

$$\gamma_{xz}^{\Delta R_1} = (V_{pp\pi} - V_{pp\sigma})\sin\theta\cos\theta,$$

$$\gamma_{xz}^{\Delta R_2} = \gamma_{xz}^{\Delta R3} = -\frac{1}{2}(V_{pp\pi} - V_{pp\sigma})\sin\theta\cos\theta,$$

$$\gamma_{xs}^{\Delta R_1} = V_{sp\sigma}\sin\theta,$$

$$\gamma_{xs}^{\Delta R_2} = \gamma_{xs}^{\Delta R3} = -\frac{1}{2}V_{sp\sigma}\sin\theta,$$

$$\gamma_{yy}^{\Delta R_1} = V_{pp\pi}, \tag{2.26}$$

$$\gamma_{yy}^{\Delta R_2} = \gamma_{yy}^{\Delta R_3} = V_{pp\pi} + \frac{3}{4}(V_{pp\pi} - V_{pp\sigma})\sin^2\theta,$$

$$\gamma_{yz}^{\Delta R_1} = 0,$$

$$\gamma_{yz}^{\Delta R_2} = -\gamma_{ys}^{\Delta R_3} = \frac{\sqrt{3}}{2}(V_{pp\sigma} - V_{pp\pi})\sin\theta\cos\theta,$$

$$\gamma_{ys}^{\Delta R_1} = 0,$$

$$\gamma_{ys}^{\Delta R_2} = -\gamma_{ys}^{\Delta R_3} = -\frac{\sqrt{3}}{2}V_{sp\sigma}\sin\theta,$$

$$\gamma_{zz}^{\Delta R_1} = \gamma_{zz}^{\Delta R_2} = \gamma_{zz}^{\Delta R_3} = V_{pp\pi} + (V_{pp\sigma} - V_{pp\pi})\cos^2\theta,$$

$$\gamma_{zs}^{\Delta R_1} = \gamma_{zs}^{\Delta R_2} = \gamma_{zs}^{\Delta R_3} = -V_{sp\sigma}\cos\theta;$$

$$\gamma_{ss}^{\Delta R_1} = \gamma_{ss}^{\Delta R_2} = \gamma_{ss}^{\Delta R_3} = V_{ss\sigma}.$$

It should be noted that the strong sp^3 hybridization not only causes the hoppings between $5p_z$ and $(5p_x, 5p_y, 5s)$ orbitals, e.g. $\gamma_{zx}^{\Delta R_1} = (V_{pp\pi} - V_{pp\sigma})\sin\theta\cos\theta$, but also results in the non-parallel misorientation of $p\pi$ orbitals $(\gamma_{zz}^{\Delta R_J} = V_{pp\pi}\sin^2\theta + V_{pp\sigma}\cos^2\theta)$. The SOC interaction under the Coulomb central potential, $V_{SOC} = \lambda_{SOC}\mathbf{L}\cdot\mathbf{S}$, can also be expressed as

$$V_{SOC} = \lambda_{SOC}\left(\frac{L_+s_- + L_-s_+}{2} + L_zs_z\right), \tag{2.27}$$

where $L_\pm = L_x \pm iL_y$ $(s_\pm = s_x \pm is_y)$ is the ladder operator for the angular momentum (spin). Such operators will play an important role in the spin-dependent Hamiltonian. The 5p$_z$ orbital and the composite orbitals of $\frac{1}{\sqrt{2}}$ $(5p_x \pm i5p_y)$ correspond to the magnetic quantum number of 0 and ± 1, respectively. Therefore, the SOC between the

$5p_z$ and $(5p_x, 5p_y)$ orbitals leads to the splitting of states and an interchange of spin configurations, whereas that between the $5p_x$ and $5p_y$ orbitals causes the splitting of states with opposite spin configurations. In the calculations, only the intra-atomic V_{SOC} is considered and the related matrix elements are

$$\langle A_{p_\alpha, m} | V_{SOC} | A_{p_\beta, m'} \rangle = \langle B_{p_\alpha, m} | V_{SOC} | B_{p_\beta, m'} \rangle = \frac{\lambda_{SOC}}{2}(-i\varepsilon_{\alpha\beta\gamma}\sigma^\gamma_{mm'}), \qquad (2.28)$$

where (α, β, γ) denote the (x, y, z) directions, ε is the permutation operator, and σ the Pauli spin matrix.

In the Landau gauge $\mathbf{A} = (0, B_z x, 0)$, an enlarged rectangle includes $4R_B$ ($4 \times 21665/ B_z$; B_z in unit of T) Sn atoms; furthermore, four orbitals and two spin configurations need to be taken into consideration. As a result, the magnetic Hamiltonian is the linear superposition of the $32R_B$ tight-binding functions $\{|A_{Jom}\rangle; |B_{Jom}\rangle | J = 1, 2, ..., 2R_B; o = 5p_x, 5p_y, 5p_z, 5s; m = \uparrow, \downarrow\}$. The independent Hamiltonian matrix elements are characterized by

$$\langle A_{Jom} | H | B_{Io'm} \rangle = \left(\gamma^{\Delta \mathbf{R}_1}_{oo'} t_1 + \gamma^{\Delta \mathbf{R}_2}_{oo'} t_2\right)\delta_{J,I} + \gamma^{\Delta \mathbf{R}_3}_{oo'} t_3 \delta_{J,I+1}, \qquad (2.29)$$

where the three position-dependent phases, t_1, t_2, and t_3, are similar to those of monolayer graphene (equation (2.5)). Most of the matrix elements are complex numbers even for the $(k_x = 0, k_y = 0)$ LL state, in contrast with those of layered graphenes. Numerical calculations will become more difficult in multi-orbital bonding systems. A perpendicular electric field in buckled monolayer tinene will dramatically change electronic properties. Moreover, similar calculations, being sensitive to the strength of orbital hybridization and SOC, could also be made for silicene and germanene.

2.4 Monolayer and bilayer phosphorenes

Monolayer phosphorene, with a puckered honeycomb structure, has a primitive unit cell containing four phosphorus atoms, as indicated by the dashed green lines in figure 2.3(a). Two of the four phosphorus atoms are located on the lower (magenta circles) or upper (blue circles) sublattice sites. The two bond lengths for in-plane and inter-plane P-P connections are $b_1 = 2.24$ Å and $b_2 = 2.22$ Å, respectively. The angle between two in-plane bonds is $\alpha' = 100.7°$, and that between the inter-plane bond and the x-axis is $\beta' = 69.2°$. The lattice vectors are $\mathbf{a}_1 = 2b_1\sin(\frac{\alpha'}{2})\hat{y}$ and $\mathbf{a}_2 = [b_2\cos(\beta') + 2b_1\cos(\frac{\alpha'}{2})]\hat{x}$, as indicated by the green arrows in figure 2.3(a). For bilayer phosphorene, AB stacking is the most stable configuration, as revealed in black phosphorus. Such stacking could be regarded as shifting the bottom layer by half of the cell along either $\mathbf{a}_1$ or $\mathbf{a}_2$ direction. As a result, the edge corresponds to the center for the puckered hexagons in two neighboring layers, or vice versa.

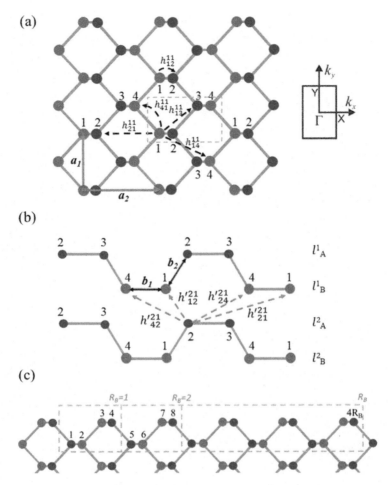

Figure 2.3. Geometric structures of (a) monolayer and (b) bilayer phosphorenes, respectively, for the top and side views with the intralayer and interlayer hopping integrals. (c) The magnetically enlarged unit cell of monolayer system.

The low-lying electronic structure is mainly from the $3p_z$-orbital hybridizations. The Hamiltonian of monolayer phosphorene (bilayer AB stacking) is a 4×4 (8×8) Hermitian matrix. The significant atomic interactions include the position-dependent five intralayer h_{IJ}^{ll} and four interlayer $h_{IJ}^{ll'}$ ($l \neq l'$) hopping integrals. The former, as clearly shown by the dashed black arrows in figure 2.3(a), are $h_{12}^{11} = 3.665$ eV, $h_{21}^{11} = -0.055$ eV, $h_{13}^{11} = -0.105$ eV, $h_{14}^{11} = -0.205$ eV, and $h_{41}^{11} = -1.22$ eV. The latter (figure 2.3(b)) are $h_{12}^{21} = 0.295$ eV, $h_{21}^{21} = -0.091$ eV, $h_{24}^{21} = -0.151$ eV, and $h_{42}^{21} = 0.273$ eV. The matrix elements related to the intralayer interactions are given by:

$$\langle B_{I\mathbf{k}}^{l}|H|A_{J\mathbf{k}}^{l'}\rangle = \langle A_{I\mathbf{k}}^{l}|H|B_{J\mathbf{k}}^{l'}\rangle = (h_{12}^{11}t_2' + h_{21}^{11}t_5')\delta_{I+1,J}\delta_{l,l'},$$

$$\langle B_{I\mathbf{k}}^{l}|H|A_{J\mathbf{k}}^{l'}\rangle = \langle A_{I\mathbf{k}}^{l}|H|B_{J\mathbf{k}}^{l'}\rangle = (h_{13}^{11}t_4' + h_{13}^{11}tt_4')\delta_{I+2,J}\delta_{l,l'},$$

$$\langle B_{I\mathbf{k}}^{l}|H|B_{J\mathbf{k}}^{l'}\rangle = (h_{41}^{11}tt_1' + h_{14}^{11}t_3')\delta_{I+3,J}\delta_{l,l'},$$

$$\langle B_{I\mathbf{k}}^{l}|H|B_{J\mathbf{k}}^{l'}\rangle = (h_{41}^{11}t_1' + h_{14}^{11}t_3')\delta_{I,J+3}\delta_{l,l'}, \qquad (2.30)$$

$$\langle A_{I\mathbf{k}}^{l}|H|A_{J\mathbf{k}}^{l'}\rangle = (h_{41}^{11}tt_1' + h_{14}^{11}t_3')\delta_{I,J-1}\delta_{l,l'},$$

$$\langle A_{I\mathbf{k}}^{l}|H|A_{J\mathbf{k}}^{l'}\rangle = (h_{41}^{11}tt_1' + h_{14}^{11}tt_3')\delta_{I+1,J}\delta_{l,l'},$$

where $I = 1, 2, 3, 4$. The position-dependent phase terms for the intralayer hoppings are written below:

$$t_1' = \exp i(k_x d_{1x} + k_y d_{1y}) + \exp i(k_x d_{1x} - k_y d_{1y}),$$

$$tt_1' = \exp i(-k_x d_{1x} + k_y d_{1y}) + \exp i(-k_x d_{1x} - k_y d_{1y}),$$

$$t_2' = \exp i(k_x d_{2x}),$$

$$t_3' = \exp i(k_x d_{3x} + k_y d_{1y}) + \exp i(k_x d_{3x} - k_y d_{1y}),$$

$$tt_3' = \exp i(-k_x d_{3x} + k_y d_{1y}) + \exp i(-k_x d_{3x} - k_y d_{1y}), \qquad (2.31)$$

$$t_4' = \exp i(k_x d_{4x} + k_y d_{1y}) + \exp i(k_x d_{4x} - k_y d_{1y}),$$

$$tt_4' = \exp i(-k_x d_{4x} + k_y d_{1y}) + \exp i(-k_x d_{4x} - k_y d_{1y}),$$

$$t_5' = \exp i(-k_x d_{5x}),$$

where d_{1x-5x} (d_{1y}) is the corresponding distance along the x (y)-axis between the interacting lattice sites. These distance parameters are expressed below:

$$d_{1x} = b_1 \cos\left(\frac{\alpha'}{2}\right),$$

$$d_{2x} = b_2 \cos\beta',$$

$$d_{3x} = b_1 \cos\left(\frac{\alpha'}{2}\right) + 2b_2 \cos\beta',$$

$$d_{4x} = b_1 \cos\left(\frac{\alpha'}{2}\right) + b_2 \cos\beta', \qquad (2.32)$$

$$d_{5x} = 2b_1 \cos\left(\frac{\alpha'}{2}\right) + b_2 \cos\beta';$$

$$d_{1y} = b_1 \sin\left(\frac{\alpha'}{2}\right).$$

As for interlayer interactions, the independent Hamiltonian matrix elements are:

$$\langle A_{I\mathbf{k}}^{l}|H|B_{J\mathbf{k}}^{l'}\rangle = (h_{42}^{21}tt_2'' + h_{42}^{21}t_2'' + h_{24}^{21}tt_3'' + h_{24}^{21}t_3'')\delta_{I+1,J}\delta_{l,l'+1},$$

$$\langle A_{I\mathbf{k}}^{l}|H|B_{J\mathbf{k}}^{l'}\rangle = (h_{42}^{21}tt_2'' + h_{42}^{21}t_2'' + h_{24}^{21}tt_3'' + h_{24}^{21}t_3'')\delta_{I,J+2}\delta_{l,l'+1},$$

$$\langle A_{I\mathbf{k}}^{l}|H|B_{J\mathbf{k}}^{l'}\rangle = (h_{12}^{21}tt_1'' + h_{21}^{21}t_4''')\delta_{I,J+1}\delta_{l,l'+1}; \qquad (2.33)$$

$$\langle A_{I\mathbf{k}}^{l}|H|B_{J\mathbf{k}}^{l'}\rangle = (h_{12}^{21}t_1'' + h_{21}^{21}tt_4''')\delta_{I+1,J}\delta_{l,l'+1}.$$

The position-dependent phase terms for the interlayer hoppings are as follows:

$$\begin{aligned}
t_1'' &= \exp i(k_x d_{2x} + k_y d_{1y}) + \exp i(k_x d_{2x} - k_y d_{1y}), \\
tt_1'' &= \exp i(-k_x d_{2x} + k_y d_{1y}) + \exp i(-k_x d_{2x} - k_y d_{1y}), \\
t_2'' &= \exp i(k_x d_{4x}), \\
tt_2'' &= \exp i(-k_x d_{4x}), \\
t_3'' &= \exp i(k_x d_{4x} + k_y 2d_{1y}) + \exp i(k_x d_{4x} - k_y 2d_{1y}), \\
tt_3'' &= \exp i(-k_x d_{4x} + k_y 2d_{1y}) + \exp i(-k_x d_{4x} - k_y 2d_{1y}), \\
t_4'' &= \exp i(k_x d_{5x} + k_y d_{1y}) + \exp i(k_x t_{5x} - k_y d_{1y}); \\
tt_4'' &= \exp i(-k_x d_{5x} + k_y d_{1y}) + \exp i(-k_x d_{5x} - k_y d_{1y}).
\end{aligned} \tag{2.34}$$

When monolayer and bilayer phosphorene systems exist in a perpendicular magnetic field, the magnetic flux through a puckered hexagon is $\Phi = a_1 a_2 B_z / 2$. The vector potential, $\mathbf{A} = (B_z x) \hat{y}$, leads to a new period along $\hat{x}$ and thus an enlarged rectangular unit cell with $4 R_B = 4\Phi_0 / \Phi$ atoms in monolayer, as illustrated in figure 2.3(c). The reduced first Brillouin zone has an area of $4\pi^2 / a_1 a_2 R_B$. For bilayer phosphorene, the magnetic Hamiltonian matrix is very large with $8 R_B \times 8 R_B$ matrix elements within achievable experimental field strengths, e.g. the dimension of 16 800 at $B_z = 30$ T. The magnetic matrix elements can be obtained by replacing the phase terms in equation (2.31) by those with extra Peierls phases, as given below:

(I) intralayer interactions:

$$\begin{aligned}
tt'_{1,I=\text{odd}} &= \exp i\left(-k_x d_{1x} + k_y d_{1y} + \pi \frac{\Phi}{\Phi_0}\left(I - 1 + \frac{2d_{2x} - d_{3x}}{d_{4x}}\right)\right) \\
&\quad + \exp i\left(k_x d_{1x} - k_y d_{1y} - \pi \frac{\Phi}{\Phi_0}\left(I - 1 + \frac{2d_{2x} - d_{3x}}{2d_{4x}}\right)\right), \\
t'_{3,I=\text{odd}} &= \exp i\left(k_x d_{3x} + k_y d_{1y} + \pi \frac{\Phi}{\Phi_0}\left(I - 1 + \frac{2d_{2x} + d_{1x}}{2d_{4x}}\right)\right) \\
&\quad + \exp i\left(k_x d_{3x} - k_y d_{1y} + \pi \frac{\Phi}{\Phi_0}\left(I - 1 + \frac{2d_{2x} + d_{1x}}{2d_{4x}}\right)\right), \\
tt'_{3,I=\text{odd}} &= \exp i\left(-k_x d_{3x} + k_y d_{1y} + \pi \frac{\Phi}{\Phi_0}\left(I - 1 - \frac{2d_{2x} + d_{1x}}{2d_{4x}}\right)\right) \\
&\quad + \exp i\left(-k_x d_{3x} - k_y d_{1y} - \pi \frac{\Phi}{\Phi_0}\left(I - 1 - \frac{2d_{2x} + d_{1x}}{2d_{4x}}\right)\right), \\
t'_{4,I=\text{odd}} &= \exp i\left(k_x d_{4x} + k_y d_{1y} + \pi \frac{\Phi}{\Phi_0}\left(I - \frac{1}{2}\right)\right) \\
&\quad + \exp i\left(k_x d_{4x} - k_y d_{1y} - \pi \frac{\Phi}{\Phi_0}\left(I - \frac{1}{2}\right)\right),
\end{aligned} \tag{2.35}$$

$$tt'_{4,I=\text{odd}} = \exp i\left(-k_x d_{4x} + k_y d_{1y} + \pi\frac{\Phi}{\Phi_0}\left(I - \frac{3}{2}\right)\right)$$
$$+ \exp i\left(-k_x d_{4x} - k_y d_{1y} - \pi\frac{\Phi}{\Phi_0}\left(I - \frac{3}{2}\right)\right),$$

$$t'_{4,I=\text{even}} = \exp i\left(k_x d_{4x} + k_y d_{1y} + \pi\frac{\Phi}{\Phi_0}\left(I - 1 + \frac{2d_{2x} + d_{4x}}{2d_{4x}}\right)\right)$$
$$+ \exp i\left(k_x d_{4x} - k_y d_{1y} - \pi\frac{\Phi}{\Phi_0}\left(I - 1 + \frac{2d_{2x} + d_{4x}}{2d_{4x}}\right)\right),$$

$$tt'_{4,I=\text{even}} = \exp i\left(-k_x d_{4x} + k_y d_{1y} + \pi\frac{\Phi}{\Phi_0}\left(I - 1 - \frac{2d_{2x} + d_{4x}}{2d_{4x}}\right)\right)$$
$$+ \exp i\left(-k_x d_{4x} - k_y d_{1y} - \pi\frac{\Phi}{\Phi_0}\left(I - 1 - \frac{2d_{2x} + d_{4x}}{2d_{4x}}\right)\right), \qquad (2.36)$$

$$t'_{1,I=\text{even}} = \exp i\left(k_x d_{1x} + k_y d_{1y} + \pi\frac{\Phi}{\Phi_0}\left(I - 1 + \frac{2d_{2x} + d_{1x}}{2d_{4x}}\right)\right)$$
$$+ \exp i\left(k_x d_{1x} - k_y d_{1y} - \pi\frac{\Phi}{\Phi_0}\left(I - 1 + \frac{2d_{2x} + d_{1x}}{2d_{4x}}\right)\right);$$

$$tt'_{3,I=\text{even}} = \exp i\left(-k_x d_{3x} + k_y d_{1y} + \pi\frac{\Phi}{\Phi_0}\left(I - 1 + \frac{2d_{2x} - d_{3x}}{2d_{4x}}\right)\right)$$
$$+ \exp i\left(-k_x d_{3x} - k_y d_{3y} - \pi\frac{\Phi}{\Phi_0}\left(I - 1 + \frac{2d_{2x} - d_{3x}}{2d_{4x}}\right)\right).$$

(II) interlayer interactions:

$$t''_{1,I=\text{odd}} = \exp i\left(k_x d_{2x} + k_y d_{1y} + \pi\frac{\Phi}{\Phi_0}\left(I - 1 + \frac{d_{2x}}{2d_{4x}}\right)\right)$$
$$+ \exp i\left(k_x d_{2x} - k_y d_{1y} - \pi\frac{\Phi}{\Phi_0}\left(I - 1 + \frac{d_{2x}}{2d_{4x}}\right)\right),$$

$$tt''_{1,I=\text{odd}} = \exp i\left(-k_x d_{2x} + k_y d_{1y} + \pi\frac{\Phi}{\Phi_0}\left(I - 1 - \frac{d_{2x}}{2d_{4x}}\right)\right)$$
$$+ \exp i\left(-k_x d_{2x} - k_y d_{1y} - \pi\frac{\Phi}{\Phi_0}\left(I - 1 - \frac{d_{2x}}{2d_{4x}}\right)\right),$$

$$t''_{1,I=\text{even}} = \exp i\left(k_x d_{2x} + k_y d_{1y} + \pi\frac{\Phi}{\Phi_0}\left(I - 1 - \frac{d_{2x}}{2d_{4x}}\right)\right)$$
$$+ \exp i\left(k_x d_{2x} - k_y d_{1y} - \pi\frac{\Phi}{\Phi_0}\left(I - 1 - \frac{d_{2x}}{2d_{4x}}\right)\right),$$

$$tt''_{1,I=\text{even}} = \exp i\left(-k_x d_{2x} + k_y d_{1y} + \pi\frac{\Phi}{\Phi_0}\left(I - 1 + \frac{d_{2x}}{2d_{4x}}\right)\right)$$
$$+ \exp i\left(-k_x d_{2x} - k_y d_{1y} - \pi\frac{\Phi}{\Phi_0}\left(I - 1 + \frac{d_{2x}}{2d_{4x}}\right)\right),$$

$$t''_{3,I=\text{odd}} = \exp i\left(k_x d_{4x} + k_y 2d_{1y} + 2\pi\frac{\Phi}{\Phi_0}\left(I - \frac{1}{2}\right)\right)$$
$$+ \exp i\left(k_x d_{4x} - k_y 2d_{1y} - 2\pi\frac{\Phi}{\Phi_0}\left(I - \frac{1}{2}\right)\right),$$

$$tt''_{3,I=\text{odd}} = \exp i\left(-k_x d_{4x} + k_y 2d_{1y} + 2\pi\frac{\Phi}{\Phi_0}\left(I - \frac{3}{2}\right)\right)$$
$$+ \exp i\left(-k_x d_{4x} - k_y 2d_{1y} - 2\pi\frac{\Phi}{\Phi_0}\left(I - \frac{3}{2}\right)\right),$$

$$t''_{3,I=\text{even}} = \exp i\left(k_x d_{4x} + k_y 2d_{1y} + 2\pi\frac{\Phi}{\Phi_0}\left(I - 1 + \frac{2d_{2x} + d_{4x}}{2d_{4x}}\right)\right)$$
$$+ \exp i\left(k_x d_{4x} - k_y 2d_{1y} - 2\pi\frac{\Phi}{\Phi_0}\left(I - 1 + \frac{2d_{2x} + d_{4x}}{2d_{4x}}\right)\right), \qquad (2.37)$$

$$tt''_{3,I=\text{even}} = \exp i\left(-k_x d_{4x} + k_y 2d_{1y} + 2\pi\frac{\Phi}{\Phi_0}\left(I - 1 - \frac{2d_{2x} + d_{4x}}{2d_{4x}}\right)\right)$$
$$+ \exp i\left(-k_x d_{4x} - k_y 2d_{1y} - 2\pi\frac{\Phi}{\Phi_0}\left(I - 1 - \frac{2d_{2x} + d_{4x}}{2d_{4x}}\right)\right),$$

$$t''_{4,I=\text{odd}} = \exp i\left(k_x d_{5x} + k_y d_{1y} + \pi\frac{\Phi}{\Phi_0}\left(I - 1 - \frac{d_{5x}}{2d_{4x}}\right)\right)$$
$$+ \exp i\left(k_x d_{5x} - k_y d_{1y} - \pi\frac{\Phi}{\Phi_0}\left(I - 1 - \frac{d_{5x}}{2d_{4x}}\right)\right),$$

$$tt''_{4,I=\text{odd}} = \exp i\left(-k_x d_{5x} + k_y d_{1y} + \pi\frac{\Phi}{\Phi_0}\left(I - 1 + \frac{d_{5x}}{2d_{4x}}\right)\right)$$
$$+ \exp i\left(-k_x d_{5x} - k_y d_{1y} - \pi\frac{\Phi}{\Phi_0}\left(I - 1 + \frac{d_{5x}}{2d_{4x}}\right)\right),$$

$$t''_{4,I=\text{even}} = \exp i\left(k_x d_{5x} + k_y d_{1y} + \pi\frac{\Phi}{\Phi_0}\left(I - 1 + \frac{d_{2x}}{2d_{4x}}\right)\right)$$
$$+ \exp i\left(k_x d_{5x} - k_y d_{1y} - \pi\frac{\Phi}{\Phi_0}\left(I - 1 + \frac{d_{2x}}{2d_{4x}}\right)\right),$$

$$tt''_{4,I=\text{even}} = \exp i\left(-k_x d_{5x} + k_y d_{1y} + \pi\frac{\Phi}{\Phi_0}\left(I - 1 - \frac{d_{2x}}{2d_{4x}}\right)\right)$$
$$+ \exp i\left(-k_x d_{5x} - k_y d_{1y} - \pi\frac{\Phi}{\Phi_0}\left(I - 1 - \frac{d_{2x}}{2d_{4x}}\right)\right), \qquad (2.38)$$

where $I = 1, 2, 3..., 4R_B$. It should be noted that the magnetic Hamiltonian matrix elements are real numbers for the $(k_x = 0, k_y)$ LL states, as revealed in layered graphenes.

2.5 MoS$_2$

A MoS$_2$ monolayer, as shown in figures 2.4(a) and (b), is composed of three atomic layers. A single layer of molybdenum atoms is sandwiched by two sulfur layers, in which the former alone forms a 2D triangular lattice. Each Mo atom interacts with six neighboring Mo atoms, as well as six neighboring S atoms (three on the top/bottom layer). The outermost shells of the Mo atom are 4d orbitals and those of the S atom are 3p orbitals. According to the theoretical calculations [23–25], the low-lying energy bands are mainly determined by three $(4d_{z^2}, 4d_{xy}, 4d_{x^2-y^2})$ orbitals of the molybdenum atoms, i.e. the valence and conduction bands nearest to the chemical potential are predominantly contributed to from such orbitals. In contrast, energy bands dominated by the $(4d_{xz}, 4d_{yz})$ orbitals of molybdenum, and the $(3p_x, 3p_y, 3p_z)$ orbitals of sulfide, as well as other inner orbitals, belong to the higher/deeper electronic states, so that these orbitals hardly affect the magnetic quantization.

Without applying a magnetic field, a unit cell of parallelogram (blue dashed curves) contains one Mo atom with six nearest-neighbor interactions. The

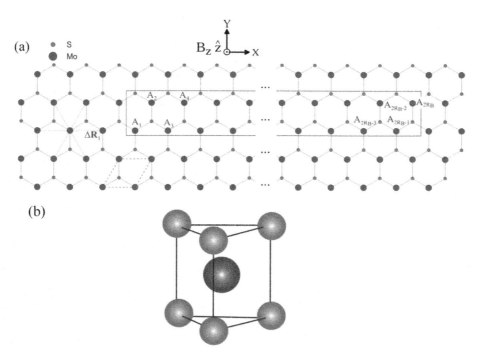

Figure 2.4. (a) Geometric structures for a MoS$_2$ monolayer with an enlarged rectangular unit cell in $\mathbf{B}_z\hat{z}$ and (b) the structure of trigonal prismatic coordination.

independent Hamiltonian matrix elements, which are built from the three tight-binding functions of $(4d_{z^2}, 4d_{xy}, 4d_{x^2-y^2})$ orbitals, present the analytic form

$$\langle 4d_{z^2}|H|4d_{z^2}\rangle = 2\gamma_{11}(\cos 2\alpha'' + 2\cos\alpha''\cos\alpha''') + E_{o1},$$

$$\langle 4d_{z^2}|H|4d_{xy}\rangle = -2\sqrt{3}\gamma_{13}\sin\alpha''\sin\alpha'''$$
$$+ 2i\gamma_{12}(\sin 2\alpha'' + \sin\alpha''\cos\alpha'''),$$

$$\langle 4d_{z^2}|H|4d_{x^2-y^2}\rangle = 2\sqrt{3}\gamma_{12}\cos\alpha''\sin\alpha'''$$
$$+ 2\gamma_{13}(\cos 2\alpha'' - \cos\alpha''\cos\alpha'''), \qquad (2.39)$$

$$\langle 4d_{xy}|H|4d_{xy}\rangle = 2\gamma_{22}\cos 2\alpha'' + (\gamma_{22} + 3\gamma_{33})\cos\alpha''\cos\alpha'''$$
$$+ E_{o2},$$

$$\langle 4d_{xy}|H|4d_{x^2-y^2}\rangle = \sqrt{3}(\gamma_{33} - \gamma_{22})\sin\alpha''\sin\alpha'''$$
$$+4i\gamma_{23}\sin\alpha''(\cos\alpha'' - \cos\alpha'''),$$

$$\langle 4d_{x^2-y^2}|H|4d_{x^2-y^2}\rangle = 2\gamma_{33}\cos 2\alpha'' + (3\gamma_{22} + \gamma_{33})\cos\alpha''\cos\alpha''' + E_{o2},$$

where $\alpha'' = \frac{1}{2}k_x a$ and $\alpha''' = \frac{\sqrt{3}}{2}k_y a$. $a = 3.19$ Å is the lattice constant between two neighboring Mo atoms. The nearest-neighbor hopping integrals of atomic orbitals are chosen as $\gamma_{11} = -0.184$ eV $(\chi_{d_{z^2}}(\mathbf{r}_I), \chi_{d_{z^2}}(\mathbf{r}_I + \Delta\mathbf{R}_1))$, $\gamma_{12} = 0.401$ eV $(\chi_{d_{z^2}}(\mathbf{r}_I), \chi_{d_{xy}}(\mathbf{r}_I + \Delta\mathbf{R}_1))$, $\gamma_{13} = 0.507$ eV $(\chi_{d_{z^2}}(\mathbf{r}_I), \chi_{d_{x^2-y^2}}(\mathbf{r}_I + \Delta\mathbf{R}_1))$, $\gamma_{22} = 0.218$ eV $(\chi_{d_{xy}}(\mathbf{r}_I), \chi_{d_{xy}}(\mathbf{r}_I + \Delta\mathbf{R}_1))$, $\gamma_{23} = 0.057$ eV $(\chi_{d_{xy}}(\mathbf{r}_I), \chi_{d_{x^2-y^2}}(\mathbf{r}_I + \Delta\mathbf{R}_1))$, and $\gamma_{33} = 0.388$ eV $(\chi_{d_{x^2-y^2}}(\mathbf{r}_I), \chi_{d_{x^2-y^2}}(\mathbf{r}_I + \Delta\mathbf{R}_1))$, where $\chi_{d_{z^2}}(\mathbf{r}_I)$, $\chi_{d_{xy}}(\mathbf{r}_I)$ and $\chi_{d_{x^2-y^2}}(\mathbf{r}_I)$ are the atomic orbitals centered at $\mathbf{r}_I$ [26]. The on-site energy is $E_{o1} = 1.046$ eV added to the $|4d_{z^2}\rangle$ orbital and $E_{o2} = 2.104$ eV added to the $|4d_{xy}\rangle$ and $|4d_{x^2-y^2}\rangle$ orbitals.

Considering the spin degree of freedom, the number of the basis is doubled as $\{|4d_{z^2}, \uparrow\rangle, |4d_{xy}, \uparrow\rangle, |4d_{x^2-y^2}, \uparrow\rangle, |4d_{z^2}, \downarrow\rangle, |4d_{xy}, \downarrow\rangle, |4d_{x^2-y^2}, \downarrow\rangle\}$. The intra-atomic SOC interaction, $V_{SOC} = \frac{\lambda_{SOC}}{2}L_z s_z$, is expressed as

$$V_{SOC} = \begin{bmatrix} 0 & 0 & 0 & 0 & 0 & 0 \\ 0 & 0 & 2i & 0 & 0 & 0 \\ 0 & -2i & 0 & 0 & 0 & 0 \\ 0 & 0 & 0 & 0 & 0 & 0 \\ 0 & 0 & 0 & 0 & 0 & -2i \\ 0 & 0 & 0 & 0 & 2i & 0 \end{bmatrix}, \qquad (2.40)$$

and the SOC coupling strength $\lambda_{SOC} = 0.073$ eV [26]. Since the $4d_{z^2}$ and the composite orbitals of $\frac{1}{\sqrt{2}}(4d_{x^2-y^2} \pm 4d_{xy})$ correspond to the magnetic quantum numbers of 0 and ± 2, respectively, the SOC interactions occur between the $4d_{xy}$ and $4d_{x^2-y^2}$ orbital electrons with the same spin orientation. In sharp contrast with tinene (equation (2.27)), the SOC interactions related to $\lambda_{SOC}(\frac{L_+s_- + L_-s_+}{2})$ are absent in MoS$_2$. The main reason is that $L_{\pm}$ can only result in the coupling of two orbitals with the ± 1 difference in the magnetic quantum number.

Under the influence of $B_z \hat{z}$ with $\mathbf{A} = (0, B_z x, 0)$, the unit cell becomes an enlarged rectangle containing $2R_B$ Mo atoms (figure 2.4(a)). Consequently, the wave function is the linear combination of $12R_B$ tight-binding functions $\{|o_{Im}\rangle, I = 1, 2, \ldots, 2 R_B;$ $o = 4d_{z^2}, 4d_{xy}, 4d_{x^2-y^2}; m = \uparrow, \downarrow\}$. The nonzero and independent matrix elements of the magnetic Hamiltonian, which belong to the complex numbers even at $k_x = 0$ and $k_y = 0$, are given by

$$\langle 4d_{z^2,I}|H|4d_{z^2,J}\rangle = \gamma_{11}\exp(i2\alpha'')\delta_{J,I-2} + 2\gamma_{11}\cos\beta_J\exp(i\alpha'')\delta_{J,I-1}$$
$$+ \gamma_{11}\exp(-i2\alpha'')\delta_{J,I+2}$$
$$+ 2\gamma_{11}\cos\beta_{J-1}\exp(-i\alpha'')\delta_{J,I+1} + E_o\delta_{J,I},$$

$$\langle 4d_{z^2,I}|H|4d_{xy,J}\rangle = \gamma_{12}\exp(i2\alpha'')\delta_{J,I-2}$$
$$+ [\gamma_{12}\cos\beta_J + \sqrt{3}\,i\gamma_{13}\sin\beta_J]\exp(i\alpha'')\delta_{J,I-1}$$
$$- \gamma_{12}\exp(-i2\alpha'')\delta_{J,I+2} + [-\gamma_{12}\cos\beta_{J-1}$$
$$- \sqrt{3}\,i\gamma_{13}\sin\beta_{J-1}]\exp(-i\alpha'')\delta_{J,I+1},$$

$$\langle 4d_{z^2,I}|H|4d_{x^2-y^2,J}\rangle = \gamma_{13}\exp(i2\alpha'')\delta_{J,I-2}$$
$$+ [-\gamma_{13}\cos\beta_J + \sqrt{3}\,i\gamma_{12}\sin\beta_J]\exp(i\alpha'')\delta_{J,I-1}$$
$$+ \gamma_{13}\exp(-i2\alpha'')\delta_{J,I+2} + [-\gamma_{13}\cos\beta_{J-1}$$
$$+ \sqrt{3}\,i\gamma_{12}\sin\beta_{J-1}]\exp(-i\alpha'')\delta_{J,I+1},$$

$$\langle 4d_{xy,I}|H|4d_{xy,J}\rangle = \gamma_{22}\exp(i2\alpha'')\delta_{J,I-2}$$
$$+ \left[\frac{1}{2}(\gamma_{22} + 3\gamma_{33})\cos\beta_J\right]\exp(i\alpha'')\delta_{J,I-1}$$
$$+ \gamma_{22}\exp(-i2\alpha'')\delta_{J,I+2} + \left[\frac{1}{2}(\gamma_{22} + 3\gamma_{33})\cos\beta_{J-1}\right]\exp(-i\alpha'')\delta_{J,I+1}$$
$$+ E_{o2}\delta_{J,I},$$

$$\langle 4d_{x^2-y^2,I}|H|4d_{x^2-y^2,J}\rangle = \gamma_{33}\exp(i2\alpha'')\delta_{J,I-2}$$
$$+ \left[\frac{1}{2}(3\gamma_{22} + \gamma_{33})\cos\beta_J\right]\exp(i\alpha'')\delta_{J,I-1}$$
$$+ \gamma_{33}\exp(-i2\alpha'')\delta_{J,I+2} + \left[\frac{1}{2}(3\gamma_{22} + \gamma_{33})\cos\beta_{J-1}\right]\exp(-i\alpha'')\delta_{J,I+1}$$
$$+ E_{o2}\delta_{J,I},$$

$$\langle 4d_{xy,I}|H|4d_{x^2-y^2,J}\rangle = \gamma_{23}\exp(i2\alpha'')\delta_{J,I-2} - \gamma_{23}\exp(-i2\alpha'')\delta_{J,I+2}$$

$$+ \left[-i\frac{\sqrt{3}}{2}(\gamma_{33} - \gamma_{22})\sin\beta_J - 2\gamma_{23}\cos\beta_J\right]\exp(i\alpha'')\delta_{J,I-1}$$

$$+ \left[i\frac{\sqrt{3}}{2}(\gamma_{33} - \gamma_{22})\sin\beta_{J-1} + 2\gamma_{23}\cos\beta_{J-1}\right] \tag{2.41}$$

$$\exp(-i\alpha'')\delta_{J,I+1},$$

where $\cos\beta_J = \cos[\alpha''' + \pi\dfrac{\Phi}{\Phi_0}(J + \dfrac{1}{2})]$ and $\sin\beta_J = \sin[\alpha''' + \pi\dfrac{\Phi}{\Phi_0}(J + \dfrac{1}{2})]$.

2.6 A suitable, reliable and wide-range model

The important characteristics of the generalized tight-binding model are worthy of a more detailed discussion. This model can fully comprehend the essential electronic properties under various external fields, compared with the effective-mass model. It is suitable for the uniform/modulated electric fields [27], uniform/modulated magnetic fields [28] (details in chapter 7), and the composite fields [29], since all the intrinsic interactions and the external fields are included in the calculations of band structures and wave functions simultaneously. Moreover, the dependences on the distinct sublattices, the multi- and single-orbital hybridizations and spin arrangements are explored in detail (discussed later in chapters 3–6). Such factors are responsible for the main features of the physical properties, e.g. the diverse and complicated LL anti-crossing spectra. However, the low-energy perturbation about the high symmetry points is first made on the Hamiltonian matrix elements and then the magnetic quantization is done. Such an approximation cannot accurately deal with the quantized states closely related to the partially flat, sombrero-shaped and multi-valley energy bands, as discussed above for the tetra-layer ABC-stacked graphene, monolayer tinene, and bilayer phosphorene. Evidently, further studies on the LL-induced fundamental properties will become cumbersome. For example, it is difficult to understand the magneto-optical properties and Coulomb excitations using the effective-mass model.

The generalized tight-binding model could also be used to investigate the essential physical properties of condensed-matter systems with any dimensions (details in chapter 7), e.g. various carbon-related systems. Systematic studies on the rich and unique magnetic quantization have been made for three kinds of bulk graphites (simple hexagonal, Bernal and rhombohedral graphites) [30–32], layered graphenes [33–36], 1D carbon nanotube [31], and graphene nanoribbons [38, 39]. They clearly show that a very strong competition or cooperation between the geometric structure and magnetic field can create an unusual quantization phenomenon, i.e. the magnetic quantization is greatly diversified by the different dimensions, stacking configurations, and boundary conditions. The dimensional crossover, corresponding to the dramatic changes of essential properties, is expected to be revealed in other emergent materials, e.g. layered phosphorenes and bulk phosphorous.

The calculated results are reliable at low and middle energies. The layer-, sublattice-, orbital- and spin-dependent energy spectra and wave functions can account for the other essential physical properties (chapter 7), such as quantum transport properties, optical spectra and Coulomb excitations. The generalized tight-binding model could combine with the single- and many-particle theories, when the latter are expressed in layer-dominated forms For example, the random-phase approximation (RPA), which can evaluate the effective electron–electron Coulomb interactions, needs to be transformed to obtain the layer-dominated response function [15, 16]. Up to now, theoretical studies cover optical properties [16, 17, 19, 21–23] and electronic excitations [15–19, 26, 27] of layered graphenes under electric and magnetic fields. In particular, the optical selection rules of the available transition channels and the phase diagrams of the dynamic Coulomb screenings are obtained from the main features of electronic properties. In short, the generalized tight-binding model can solve the various Hamiltonians efficiently and thus is very useful in understanding the essential physical properties of condensed-mater systems.

2.7 Numerical calculations

In this section, we introduce the generalized tight-binding model, which has been adapted to deal with different interactions in emergent 2D materials and to obtain the corresponding Hamiltonian matrices. The eigenvalues and eigenfunctions can be obtained by diagonalizing the Hamiltonian matrix. To gain an understanding of the electronic properties, the band structure in the first Brillouin zone is required; a proposition that implies the diagonalization of Hamiltonian matrices for thousands of different $k's$. In the presence of $\mathbf{B}_z$, the first Brillouin zone is greatly reduced, but the deduced magnetic Hamiltonian matrix is huge and difficult to be diagonalized. These problems can be solved by means of numerical calculations. We take monolayer graphene as an example to explain the numerical calculation process. At $B_z = 0$, there are two carbon atoms in a unit cell and the interaction between $2p_z$ orbitals of carbon atoms is considered. The 2×2 Hamiltonian matrix is expressed as

$$\begin{bmatrix} 0 & \langle A_\mathbf{k}|H|B_\mathbf{k}\rangle \\ \langle A_\mathbf{k}|H|B_\mathbf{k}\rangle^* & 0 \end{bmatrix}. \tag{2.42}$$

The k-dependent matrix elements may be obtained from equation (2.2). The Hermitian matrix can be numerically solved by using numerical libraries such as IMSL and LAPACK. The application of $\mathbf{B}z$ enlarges the unit cell and the magnetic Hamiltonian matrix. In the bases $|A_{1\mathbf{k}}\rangle$, $|B_{1\mathbf{k}}\rangle$, $|A_{2\mathbf{k}}\rangle$, $|B_{2\mathbf{k}}\rangle$, $\ldots |A_{2R_B-1\mathbf{k}}\rangle$, $|B_{2R_B-1\mathbf{k}}\rangle$, $|A_{2R_B\mathbf{k}}\rangle$; $|B_{2R_B\mathbf{k}}\rangle$, the $4R_B \times 4R_B$ Hamiltonian matrix is

$$\begin{bmatrix} 0 & p_1 & 0 & 0 & \cdots & 0 & 0 & t_3 \\ p_1^* & 0 & t_3 & 0 & \cdots & 0 & 0 & 0 \\ 0 & t_3^* & 0 & p_2 & \cdots & 0 & 0 & 0 \\ 0 & 0 & p_2^* & 0 & t_3 & \cdots & 0 & 0 \\ \vdots & \vdots & \vdots & t_3^* & \vdots & \cdots & \vdots & \vdots \\ 0 & 0 & 0 & \vdots & \cdots & \vdots & t_3 & 0 \\ 0 & 0 & 0 & 0 & \cdots & t_3^* & 0 & p_{2R_B} \\ t_3^* & 0 & 0 & 0 & \cdots & 0 & p_{2R_B}^* & 0 \end{bmatrix}. \tag{2.43}$$

In this notation, p_I is defined as $t_{1,I} + t_{2,I}$ and the elements of the magnetic Hamiltonian matrix are determined by equations (2.4) and (2.5). For small B_z, diagonalizing a $4R_B \times 4R_B$ Hamiltonian matrix is a difficult task ($R_B \sim 7900$ at $B_z = 10$ T). It is very time-consuming even on a high-performance computer with a considerable amount of physical memory. Fortunately, this problem can be simplified by rearranging the bases. In the bases $|A_{1\mathbf{k}}\rangle$, $|B_{2R_B\mathbf{k}}\rangle$, $|B_{1\mathbf{k}}\rangle$, $|A_{2R_B\mathbf{k}}\rangle$, $|A_{2\mathbf{k}}\rangle$, $|B_{2R_B-1\mathbf{k}}\rangle$, $|B_{2\mathbf{k}}\rangle$, $|A_{2R_B-1\mathbf{k}}\rangle \ldots$, the $4R_B \times 4R_B$ Hamiltonian matrix is expressed as

$$\begin{bmatrix} 0 & t_3^* & p_1 & 0 & \cdots & 0 & 0 & 0 \\ t_3 & 0 & 0 & p_{2R_B}^* & \cdots & 0 & 0 & 0 \\ p_1^* & 0 & 0 & 0 & t_3 & \cdots & 0 & 0 \\ 0 & p_{2R_B} & \vdots & 0 & \cdots & \ddots & \vdots & 0 \\ \vdots & \vdots & t_3^* & \vdots & \vdots & \vdots & p_{R_B} & \vdots \\ 0 & 0 & 0 & \ddots & \cdots & 0 & 0 & p_{R_B+1}^* \\ 0 & 0 & 0 & \cdots & p_{R_B}^* & 0 & 0 & t_3^* \\ 0 & 0 & 0 & 0 & \cdots & p_{R_B+1} & t_3 & 0 \end{bmatrix}. \tag{2.44}$$

This Hermitian matrix with an upper band width of two can be stored as a $4R_B \times 3$ band matrix. Using the library routine to solve the band matrix would substantially reduce the calculation time. The dispersionless energy spectrum means that a k state is sufficient to describe the energy spectrum of the first Brillouin zone. In addition, all of the matrix elements become real numbers at $(k_x = 0, k_y = 0)$, which can further reduce the calculation time. This numerical characteristic also applies to monolayer and bilayer phosphorenes and layered graphenes with normal stacking configurations. For silicene, germanene, tinene and MoS_2, the numerical calculation process is similar but the multi-orbital bonding and spin degree of freedom result in the enlargement of the Hamiltonian matrix. It should be noted that their Hamiltonian matrices are complex for all k's, leading to a longer calculation time.

A numerical calculation within the framework of the generalized tight-binding model enables us to thoroughly investigate the electronic properties of materials. The characteristics of electronic structures and wave functions are well depicted. Moreover, the results are accurate and reliable within a wide energy range. In this book, the energy bands along the high-symmetry directions are plotted to illustrate the main features of the electronic structure of the emergent 2D materials. The orbital decomposed state probabilities at the high-symmetry points can reveal the dominant atomic orbital of each energy. The zero-field band structure determines the main characteristics of the quantized LL states, such as the sublattice dominance in the Landau states, the field-dependent LL energies and the inter- and intra-group LL anticrossings. In the presence of B_z, the length of the enlarged unit cell is obviously greater than the magnetic length. Thus, the localized subenvelope function which is the combination of the $2R_B$ tight-binding functions exhibits a sequential spatial distribution in the enlarged unit cell. Well behaved in spatial distributions, the subenvelope functions with a normal zero-point number and a spatial symmetry/antisymmetry at the localization center can characterize the quantum numbers of the LLs.

References

[1] Charlier J-C, Michenaud J-P and Gonze X 1992 First-principles study of the electronic properties of simple hexagonal graphite *Phys. Rev.* B **46** 4531

[2] Do T N, Chang C P, Shih P H and Lin M F 2017 Stacking-enriched magnetotransport properties of few-layer graphenes *Phys. Chem. Chem. Phys.* **19** 29525

[3] Ho Y H, Chiu C W, Su W P and Lin M F 2014 Magneto-optical spectra of transition metal dichalcogenides: a comparative study *Appl. Phys. Lett.* **105** 222411

[4] Ho Y H, Wang Y H and Chen H Y 2014 Magnetoelectronic and optical properties of a MoS_2 monolayer *Phys. Rev.* B **89** 55316

[5] Huang Y C, Chang C P and Lin M F 2008 Magnetoabsorption spectra of bilayer graphene ribbons with Bernal stacking *Phys. Rev.* B **78** 115422

[6] Lin Y P, Lin C Y, Ho Y H, Do T N and Lin M F 2015 Magneto-optical properties of ABC-stacked trilayer graphene *Phys. Chem. Chem. Phys.* **17** 15921–7

[7] Chen R B, Chiu Y H and Lin M F 2014 Beating oscillations of magneto-optical spectra in simple hexagonal graphite *Comput. Phys. Commun.* **189** 60–5

[8] Chen R B, Chiu Y H and Lin M F 2012 A theoretical evaluation of the magneto-optical properties of AA-stacked graphite *Carbon* **54** 268–76

[9] Ho Y H, Chiu Y H, Lin D H, Chang C P and Lin M F 2010 Magneto-optical selection rules in bilayer Bernal graphene *ACS Nano* **4** 1465–72

[10] Wu J Y, Lin C Y, Gumbs G and Lin M F 2015 The effect of perpendicular electric field on temperature-induced plasmon excitations for intrinsic silicene *RSC Adv.* **5** 51912–8

[11] Wu J Y, Chen S C and Lin M F 2014 Temperature-dependent Coulomb excitations in silicene *New J. Phys.* **16** 125002

[12] Wu J Y, Chen S C, Gumbs G and Lin M F 2016 Feature-rich electronic excitations in external fields of 2D silicene *Phys. Rev.* B **94** 205427

[13] Wu J Y, Gumbs G and Lin M F 2014 Combined effect of stacking and magnetic field on plasmon excitations in bilayer graphene *Phys. Rev.* B **89** 165407

[14] Wu J Y, Chen S C, Roslyak O, Gumbs G and Lin M F 2011 Plasma excitations in graphene: Their spectral intensity and temperature dependence in magnetic field *ACS Nano* **5** 1026–32

[15] Ho Y H, Chang C P and Lin M F 2006 Electronic excitations of the multilayered graphite *Phys. Lett.* A **352** 446–50

[16] Ho Y H, Lu C L, Hwang C C, Chang C P and Lin M F 2006 Coulomb excitations in AA- and AB-stacked bilayer graphites *Phys. Rev.* B **74** 085406

[17] Lin M F, Chuang Y C and Wu J Y 2012 Electrically tunable plasma excitations in AA-stacked multilayer graphene *Phys. Rev.* B **86** 125434

[18] Chuang Y C, Wu J Y and Lin M F 2013 Electric field dependence of excitation spectra in AB-stacked bilayer graphene *Sci. Rep.* **3** 1368

[19] Chuang Y C, Wu J Y and Lin M F 2013 Electric-field-induced plasmon in AA-stacked bilayer graphene *Ann. Phys.* **339** 298–306

[20] Charlier J C and Michenaud J P 1992 Tight-binding density of electronic states of pregraphitic carbon *Phys. Rev.* B **46** 4540

[21] Chadi D J 1977 Spin-orbit splitting in crystalline and compositionally disordered semiconductors *Phys. Rev.* B **16** 790

[22] Cardona M, Shaklee K L and Pollak F H 1967 Electroreflectance at a semiconductor-electrolyte interface *Phys. Rev.* **154** 696

[23] Zhu Z Y, Cheng Y C and Schwingenschlögl U 2011 Giant spin-orbit-induced spin splitting in two-dimensional transition-metal dichalcogenide semiconductors *Phys. Rev.* B **84** 153402

[24] Lebègue S and Eriksson O 2009 Electronic structure of two-dimensional crystals from *ab initio* theory *Phys. Rev.* B **79** 11409

[25] Ataca C and Ciraci S 2011 Functionalization of single-layer MoS$_2$ honeycomb structures *J. Phys. Chem.* C **115** 13303

[26] Liu G-B, Shan W-Y, Yao Y, Yao W and Xiao D 2013 Three-band tight-binding model for monolayers of group-VIB transition metal dichalcogenides *Phys. Rev.* B **88** 085433

[27] Ou Y C, Chiu Y H, Lu J M, Su W P and Lin M F 2013 Electric modulation effect on magneto-optical spectrum of monolayer graphene *Comput. Phys. Commun.* **184** 1821–6

[28] Ou Y C, Sheu J K, Chiu Y H, Chen R B and Lin M F 2011 Influence of modulated fields on the Landau level properties of graphene *Phys. Rev.* B **83** 195405

[29] Ou Y C, Chiu Y H, Yang P H and Lin M F 2014 The selection rule of graphene in a composite magnetic field *Opt. Express* **22** 7473

[30] Ho Y H, Wang J, Chiu Y H, Lin M F and Su W P 2011 Characterization of Landau subbands in graphite: A tight-binding study *Phys. Rev.* B **83** 121201

[31] Ho C H, Chang C P and Lin M F 2014 Landau subband wave functions and chirality manifestation in rhombohedral graphite *Solid State Commun.* **197** 11–5

[32] Chen R B and Chiu Y H 2013 Landau subband and Landau level properties of AA-stacked graphene superlattice *J Nanosci. Nanotechnol.* **12** 2557–66

[33] Huang Y K, Chen S C, Ho Y H, Lin C Y and Lin M F 2014 Feature-rich magnetic quantization in sliding bilayer graphenes *Sci. Rep.* **4** 7509

[34] Lin C Y, Wu J Y, Chiu Y H and Lin M F 2014 Stacking-dependent magneto-electronic properties in multilayer graphenes *Phys. Rev.* B **90** 205434

[35] Lai Y H, Ho J H, Chang C P and Lin M F 2008 Magnetoelectronic properties of bilayer Bernal graphene *Phys. Rev.* B **77** 085426

[36] Lin C Y, Wu J Y, Ou Y J, Chiu Y H and Lin M F 2015 Magneto-electronic properties of multilayer graphenes *Phys. Chem. Chem. Phys.* **17** 26008–35

[37] Shyu F L, Chang C P, Chen R B, Chiu C W and Lin M F 2003 Magnetoelectronic and optical properties of carbon nanotubes *Phys. Rev.* B **67** 045405

[38] Huang Y C, Lin M F and Chang C P 2008 Landau levels and magneto-optical properties of graphene ribbons *J. Appl. Phys.* **103** 073709

[39] Huang Y C, Chang C P and Lin M F 2007 Magnetic and quantum confinement effects on electronic and optical properties of graphene ribbons *Nanotechnology* **18** 495401

Chapter 3

Graphene

The rich and unique low-lying energy bands and their magnetic quantization are thoroughly studied for few-layer graphenes, including monolayer graphene, tetra-layer ABC-stacked graphene, sliding bilayer graphene, and bilayer AB-stacked graphene. The dependences on the stacking configurations, the number of layers, and the electric and magnetic fields, are worthy of a systematic investigation, such as energy dispersions, critical points in energy-wave-vector space, special valleys, LL energy spectra, localization centers, spatial probability distributions, and van Hove singularities in DOSs. The diverse energy dispersions, the linear, parabolic, sombrero-shaped and oscillatory forms, come to exist in various graphene systems, being further reflected in the greatly diversified LLs. There exist complicated relationships between the multi-constant-energy surfaces and the quantized LLs. The focus lies in how many kinds/groups of LLs are clearly identified from the main features of the spatial oscillation modes. Specifically, the relative shift in bilayer graphene can create well-behaved, perturbed and undefined LLs, having never been observed in any other condensed-matter systems up to now. The B_z-dependent LL energy spectra present the non-crossing, crossing and anti-crossing behaviors. The critical mechanisms for them are illustrated by using the single- and multi-oscillation quantum modes. The effects due to a uniform perpendicular electric field on the LLs are discussed in detail, e g. the degeneracy splitting of the K and K′ valleys and the frequent LL anti-crossings in the E_z-dependent energy spectra. A detailed comparison between the theoretical calculations and the experimental measurement is also made.

3.1 Theoretical results

The Hamiltonian of the layered graphene, being dominated by the $2p_z$-orbital tight-binding functions in a unit cell, can be expressed as

$$H = \sum_{\langle IJ \rangle \langle ll' \rangle} - \gamma_{IJ}^{ll'} C_{Il}^{+} C_{Jl'},$$

(3.1)

doi:10.1088/978-0-7503-1674-3ch3
© IOP Publishing Ltd 2017

where $\gamma_{IJ}^{ll'}$ is the intralayer or interlayer hopping integral. C_{Il}^{+} ($C_{Jl'}$) represents the creation (annihilation) of an electron at the Ith (Jth) site of the lth (l' th) layer. A N-layer graphene has a $2N$-dimensional Hamiltonian matrix, in which it is transformed into a $4NR_B$-dimensional one in the presence of B_z. The quantized LLs are highly degenerate in the reduced first Brillouin zone. The ($k_x = 0$, $k_y = 0$) magnetic Hamiltonian, with the real matrix elements, is sufficient for calculating the energy spectra and wavefunctions. The LL state degeneracy has been examined in the numerical calculations thoroughly.

The hexagonal symmetry in monolayer graphene can create the low-lying isotropic Dirac-cone structure (figure 3.1(a)) and thus well-behaved LLs with a specific dependence on the quantum number ($n^{c,v}$) and field strength [1, 2]. As to each (k_x, k_y) state, all the LLs have eight-fold degeneracy. This comes from the equivalent K and K′valleys (figure 1.1(b)), the symmetry of $\pm B_z\hat{z}$, and the spin degree of freedom. At ($k_x = 0, k_y = 0$), the state probabilities of the degenerate LLs are localized at the 1/6, 2/6, 4/6, and 5/6 positions of the enlarged unit cell (figure 1.1(a)). The (2/6,5/6) and (1/6,4/6) states, respectively, correspond to the magnetic quantization from the K and K′ valleys [3]. The 2/6 localized LL wavefunctions, as shown in figure 3.1(d), have normal probability distributions in the oscillatory form, being identical to those of harmonic oscillators [4]. The quantum number of each LL is characterized by the number of zero points in the dominating B sublattice. The $n^{c,v} = 0$ LLs only come from the B sublattice. In general, the $n^{c,v}$ LL wavefunctions in the B sublattice are proportional to the ($n^{c,v}+1$) LL ones in the A sublattice, directly reflecting the honeycomb symmetry (the equivalence of A and B sublattices). The same features are revealed in the 1/6 case under the interchange of two sublattices. Specifically, the low-lying LL spectrum, as revealed in figure 3.1(c), is characterized by $E^{c,v} = \pm v_F\sqrt{2\hbar e n^{c,v}B_z/c}$ ($v_F = 3|\gamma_0|b/2$ the Fermi velocity), consistent with that obtained from effective-mass approximation [5, 6]. The square-root dependence is suitable at $|E^{c,v}| < 1$ eV, since the linear bands gradually change into the parabolic bands in the increment of state energy (figure 3.1(b)). In addition, the high-energy LL spectrum can also be obtained by the generalized tight-binding model [1, 2]. Specifically, the dispersionless feature of 2D LLs is dramatically changed by the distinct dimensions, e.g. the 1D quasi-LLs and the 3D Landau subbands (discussed later in chapter 7).

The main features of LLs, energy spectra, spatial distribution modes and state degeneracy, are very sensitive to the number of layers, the stacking configurations, and the perpendicular electric field. The LLs in the layered graphenes might exhibit asymmetric energy spectra at the Fermi level, a non-square-root or non-monotonous dependence on $n^{c,v}$ and B_z, and crossing or anti-crossing behaviors, mainly owing to the critical interlayer hopping integrals (e.g. those in figure 2.1(a)). Such interactions can induce three kinds of LLs with distinct distribution modes: (1) well-behaved LLs in a single mode (figure 3.1), (3.2) perturbed LLs with a main mode and side modes (discussed later in figure 3.5), and (3) undefined LLs composed of many comparable modes (figure 3.6). The LL degeneracy is reduced to half when the z → −z inversion symmetry is destroyed by the specific stacking configuration or the perpendicular

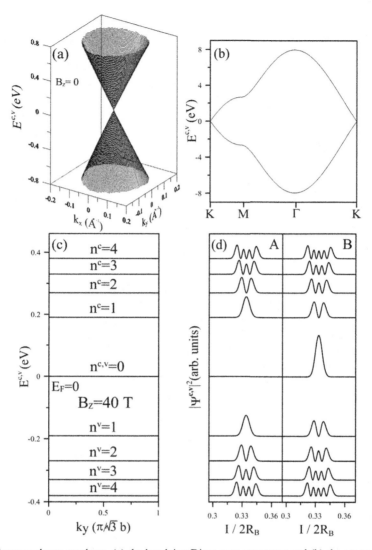

Figure 3.1. For monolayer graphene, (a) the low-lying Dirac-cone structures and (b) the energy bands along the high symmetry points. At $B_z = 40$ T, (c) the low-lying valence and conduction LLs, and (d) the probability distributions of the subenvelope functions at the A and B sublattices. The unit of the x-axis is $2R_B$, in which I represents the Ith A or B atom in an enlarged unit cell.

electric field. For example, the tri-layer AAB-stacked graphene has an obvious splitting LL spectrum with observable spacings about 10 meV [7], and a similar effect due to **E** is revealed in AB- (figure 3.7(b); [8]) and ABC-stacked graphenes [9]. In addition, the **E**-induced LL splitting is discussed later for monolayer germanene (figure 4.3), tinene and silicene (figure 4.8).

ABC-stacked tetra-layer graphene and sliding bilayer graphene are chosen to illustrate the geometry-enriched magnetic quantization. The former has four pairs of

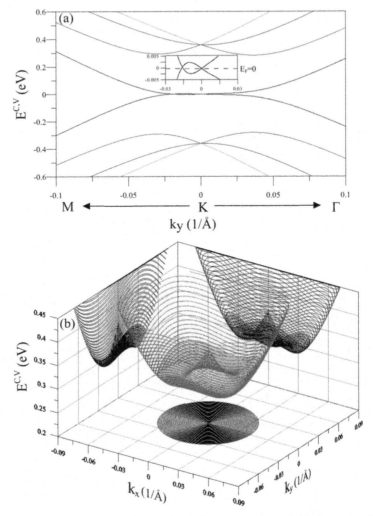

Figure 3.2. (a) Energy bands of the ABC-stacked tetralayer graphene, and (b) the conduction sombrero-shaped band in energy-wave-vector space.

valence and conduction bands, in which the first, second, third and fourth pairs, respectively, have weak, sombrero-shaped, parabolic and linear energy dispersions (black, red, blue and green curves in figure 3.2(a)). They are further quantized into the first, second, third and fourth groups of LLs (black, red, blue and green curves in figure 8(a)-8(c)), with the quantum numbers (n_1, n_2, n_3, n_4) characterized by the dominant (B^1, B^3, B^2, B^4) sublattices at the 2/6 center, respectively [10]. Apparently, the valence and the conduction LLs are asymmetric at the Fermi level. The LL energy spectrum exhibits diverse B_z-dependencies, indicating the various changes of energy bands with wave vectors (figure 3.2(a)). In general, the first group of LLs has a monotonous dependence, i.e. their energies grow with increasing B_z (figure 3.3(a)). However, the four LLs nearest to E_F, which mainly arise from the very weak energy

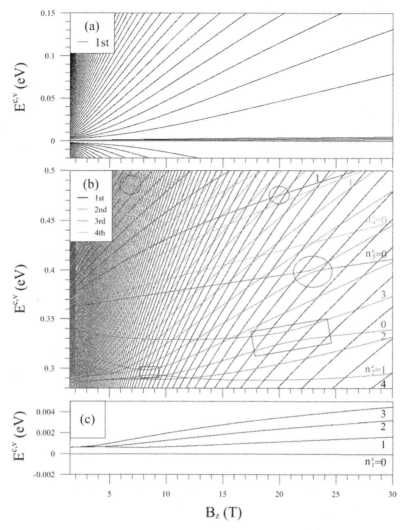

Figure 3.3. The B_z-dependent LL energy spectrum of the ABC-stacked tetralayer graphene for (a) the first group, (b) the other three groups, (c) the four LLs nearest to the Fermi levels. Rectangles and circles in (b) respectively correspond to the intragroup and intergroup anti-crossings.

dispersions dominated by the surface states (black curves near the K point in figure 3.2(a); [10]), present distribution widths smaller than 8 meV even at rather high B_z (figure 3.3(c)). These LLs localized at the two outmost graphene layers are absent in AA, AB and AAB stackings [2, 10–12].

Specifically, the second group of LLs exhibit an abnormal $n_2^{c,v}$ sequence and an unusual energy spectrum, as seen in the conduction and valence states (red curves in figure 3.3(b)). At rather small B_z, all the LLs have a reverse ordering of $E^c(n_2^c) < E^c(n_2^c - 1)$. They are initiated at a specific energy corresponding to the cusp K point of the sombrero-shaped energy band (red curve in figure 3.2(a)). This clearly illustrates that LLs are quantized from the electronic states enclosed by the inner constant-energy loop (figure 3.2(b)). With the increase of B_z, the higher-n_2^c LLs come into existence in the normal ordering since they arise from the outer constant-energy loop related to parabolic dispersions. A completely normal ordering of $E^c(n_2^c)<$ $E^c(n_2^c+1)$ is revealed only at $B_z > 100$ T, directly reflecting the fact that the electronic states under the cusp-shaped energy dispersions are only quantized into the $n_2^c = 0$ LL. the ordering of LLs is mainly determined by the competing relation between the state area covered by the cusp-shaped energy dispersions and the B_z-enhanced state degeneracy [2]. On the other hand, the third and fourth groups of LLs present a normal sequence of quantum numbers (blue and green curves in figure 3.3(b)), corresponding to the magnetic quantization of parabolic and linear energy dispersions near the K point (blue and green curves in figure 3.2(a)).

Unusual intragroup anti-crossings appear frequently in the non-monotonous LL spectrum of the second group, as illustrated in the range of 0.28 eV $< E^c(n_2^c) < 0.35$ eV by the rectangles in figure 3.3(b). In addition to a main mode, the specific interlayer hopping integrals, $(\beta_3,\beta_2,\beta_5)$ (figure 2.1(a)), cause the n_2^c LLs to possess certain side modes with zero points of $n_2^c \pm 3i$ (i is an integer) [2, 10, 13, 14]. The lower-n_2^c perturbed LLs exhibit distorted spatial distributions (figures 3.4(k)–(r)); that is, they significantly deviate from the monolayer-like single modes (figure 3.1(d)). For example, with the increase of B_z (the red circles in figure 3.4(b)), the wave functions are drastically changed during the anti-crossing of the $n_2^c = 3$ and 0 LLs, as indicated by the dominant B^3 sublattice in figure 3.4(p). When the side mode without a zero point in the $n_2^c = 3$ LL (with three zero points in the $n_2^c = 0$ LL), becomes comparable with its main mode, the same oscillation modes in these two LLs can prevent the direct crossing. Apparently, the intragroup LL anti-crossings are derived from a magnetic quantization of the non-monotonous energy bands. A similar behavior is revealed in AAB-stacked graphenes [12], while it cannot survive in the AA- and AB-stacked graphenes [2].

It should be noted that the LL anti-crossings are also present between any two distinct groups at sufficiently high B_z and $|E^{c,v}|$ [10], i.e. there exist intergroup LL anti-crossings (large circles in figure 3.3(b)). Except for the regimes of these anti-crossings, the third and the fourth groups of LL energy spectra have a normally continuous B_z-dependence. An example of the intergroup LL anti-crossing between the second and third groups is illustrated in figure 3.4(a) (red and blue circles). The cooperation of the magnetic field and interlayer hoppings can enhance the side modes of the $n_2^c = 5$ and $n_3^c = 0$ LLs and thus avoid the direct crossing within a certain B_z-range (figures 3.4(c)–(j)), especially for the B^3 and B^2 sublattices. In addition, the AB- and AAB-stacked graphenes also exhibit intergroup LL anti-crossings in the B_z-dependent energy spectra [10, 12].

In addition to well-behaved and perturbed LLs, undefined LLs are created in sliding bilayer graphene (calculation details in [3] including the interlayer hopping integrals). Specifically, the stacking configuration can be continuously changed by electrostatic-manipulation STM [15, 16]. When the configuration of bilayer graphene is transformed from AA to AB stacking by a shift ($\delta\hat{x}$) along the armchair direction (figure 3.5(a)), two vertical Dirac cones gradually become two pairs of parabolic

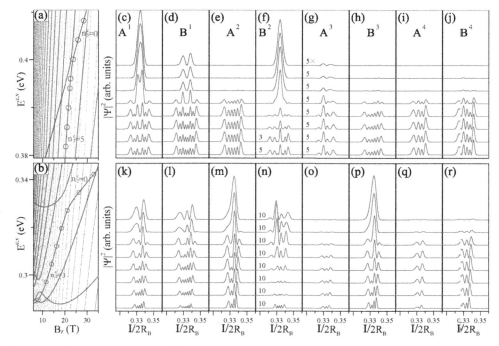

Figure 3.4. The (a) intergroup and (b) intragroup LL anti-crossing phenomena in the ABC-stacked tetralayer graphene. The spatial evolutions of subenvelope functions are respectively shown in (c)–(j) for the second and third groups of LLS, and (k)–(r) for the second group of LLs.

bands [3]. Each Dirac-cone structure, as shown in figure 3.5(b), is seriously distorted and thoroughly separated at the critical shift of ~6b/8. It is impossible to obtain the low-lying energy bands from the K-point expansion; the same is true for the LL quantization using the effective-mass approximation. The $\delta = 6b/8$ stacking exhibits eight-fold degenerate LLs, the same as monolayer graphene. However, this bilayer system has a lot of undefined LLs, as indicated in the unusual B_z-dependent energy spectrum at $| E^{c,v} | > 0.3$ eV (figure 3.5(c)). Each LL in the second group displays significant anti-crossings with all the LLs in the first group. This clearly indicates that all the LLs of the second group are composed of many comparable oscillation modes.

The undefined LLs deserve closer examination, especially in terms of the variation of field strength, as shown in figures 3.6(a) and (b) for the $\delta = 6b/8$ stacking. The spatial probability distributions of the undefined LLs exhibit relatively strong oscillations without spatial symmetry for any field strength. The number of oscillations is large even for the three initial undefined LLs of the second group (S_1, S_2;S_3 in figure 3.6(b)), and it gradually grows when B_z decreases. That is to say, they do not have a specific main mode in the entire B_z-range. As a result, the undefined LLs cannot be obtained by the low-energy perturbation method, i.e. the effective-mass approximation is not suitable for the low-lying band structures and LLs of the sliding systems. However, the main modes of the perturbed LLs can survive even in the intergroup or intragroup anti-crossings (e.g. figure 3.4).

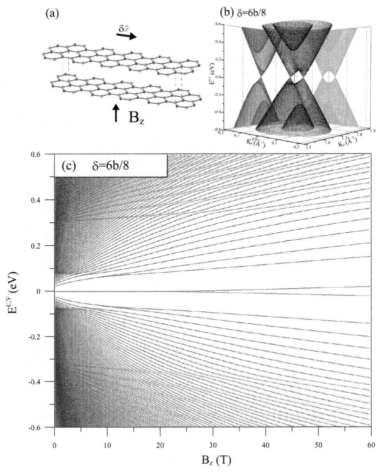

Figure 3.5. (a) Geometric structure of sliding bilayer graphene along the armchair direction ($\hat{x}$), (b) energy bands, and (c) B_z-dependent LL spectrum at $\delta = 6b/8$.

The LL splitting is clearly illustrated by a composite B_z- and E_z-field in bilayer AB stacking (figure 3.7) [8]. It is presented in layered graphenes except for high-symmetry AA-stacked systems. The AB-stacked bilayer graphene has two pairs of parabolic bands, in which the first pair exhibits a weak overlap near the Fermi level (inset in figure 3.7(a)). In general, two groups of LLs belong to well-behaved quantum modes except at very high B_z-strength. The first and the second groups are, respectively, initiated approximately at 0 eV and ± 0.4 eV (black and red curves), so that their crossings (or very few anti-crossings) occur at higher or deeper energy. A perpendicular electric field makes the four-degenerate localization centers of each LL change into two sets of equivalent localization centers; that is, (2/6, 5/6) and (1/6, 4/6), respectively, correspond to the magnetic quantization of electronic states from K and K′ valleys (solid and dashed curves in figure 3.7(b)). The LL splitting

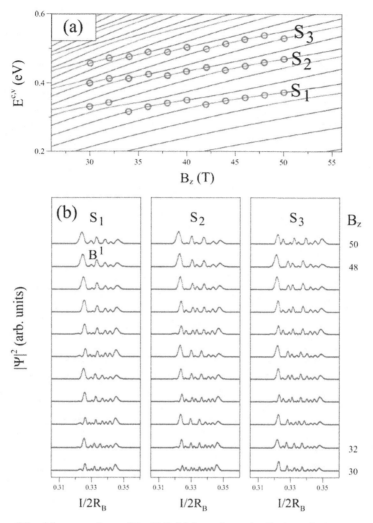

Figure 3.6. For sliding bilayer graphene of $\delta = 6b/8$, (a) the anti-crossing B_z-dependent energy spectrum of the smaller-n_2^e undefined LLs in the second group, and (b) the evolution of the probability distributions on the dominating B^1 sublattice during the intergroup anti-crossing for the initial (S_1, S_2, S_3) ones.

energy spectra exhibit monotonous E_z-dependences; furthermore, the anti-crossing behavior occurs for the same localization center. The cooperation (competition) among the interlayer atomic interactions, magnetic field and electric field can create (extinguish) the LL minor modes, leading to the LL anti-crossing (non-crossing). For example, the $n_{1K}^v = 0$ and 3 LLs of the K-dependent subgroup (the $n_{1K}^c = 0$ and 3 LLs of the K'-dependent subgroup) exhibit an obvious anti-crossing at $V_z \sim 180$– 200 meV and $B_z = 15$ T (blue circles in figure 3.7(b)), since their side modes of $n_{1K}^v \pm 3i$ ($n_{1K}^c \pm 3i$) are greatly enhanced by the important factors near the critical field. Similar changes of the main and side modes might be revealed in the B_z-dependent LL spectra of layered graphenes (e.g. tetra-layer ABC-stacked graphene in figures 3.4(a) and (b)).

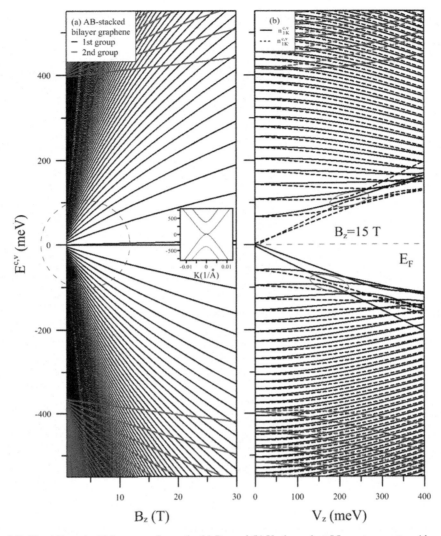

Figure 3.7. For AB-stacked bilayer graphene, the (a) B_z- and (b) V_z-dependent LL energy spectra. Also shown in the insets of (a) are zero-field band structure and linear B_z-dependence.

The DOS, which arises from two groups of LLs in bilayer AB stacking, is defined as

$$D(E) = 2 \sum_{n_1^{c,v}, n_2^{c,v}} \int_{1st\ Bz} \frac{\Gamma'}{\left[E^{c,v}\left(n^{c,v}, k_x, k_y\right) - E\right]^2 + \Gamma'2} dk_x dk_y. \tag{3.2}$$

Γ' (~0.2 meV) is a phenomenological broadening parameter used in the calculations. The factor of 2 accounts for two degenerate spin states. Similar formulas could be done for any systems in the presence or absence of external fields. The low-energy DOS is very sensitive to the field strength, as indicated in figure 3.8. At zero field

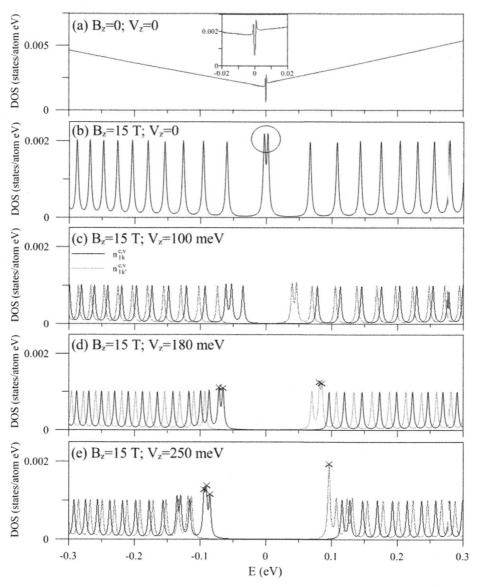

Figure 3.8. Density of states for bilayer AB stacking at (a) zero field and (b) $B_z = 10$ T combined with (c) $V_z = 50$, (d) 75 and (e) 100 meV's.

(black curve in figure 3.8(a)), only two weak shoulders, being associated with the band-edge states of the first pair of the valence and conduction bands (inset in figure 3.7(a)), respectively, occur at the right- and left-hand neighborhood of the Fermi level (inset in figure 3.8(a)). Such structures come from the extreme points in the energy-wave-vector space. Under B_z (figure 3.8(b)), they are dramatically transformed into two delta-function-like symmetric peaks due to the $n_1^v = 0$ and $n_1^c = 1$

LLs (circle in black curve); furthermore, their energy spacing is band gap. All the B_z-induced peak structures directly reflect the high-degeneracy and discrete characteristics. The peak height relies on state degeneracy and crossing behavior. It is uniform in the energy range (-0.3 eV $<$ E $<$ 0.3 eV) excluding the second group of LLs, or the intergroup crossings. The number of peak structures double in the presence of V_z (figures 3.8(c)–(e)), mainly owing to the split $n_{1K}^{c,v}$ and $n_{1K'}^{c,v}$ LLs (the solid and dashed curves). The heights of the separate symmetric peaks are reduced to half. However, DOS might present few double-peak structures with stronger intensities as a result of the LL anti-crossings or crossings (crosses).

Layered graphenes exhibit three kinds of LLs. The well-behaved, perturbed and undefined LLs are, respectively, revealed in the noncrossing/crossing, anti-crossing, and frequently anti-crossing B_z-dependent energy spectra. This behavior could be directly verified by STS measurements (details in the final paragraph of chapter 6). Moreover, the quantum modes of LLs are responsible for the magneto-optical selection rules ([3, 17]). For the well-behaved modes, the initial occupied LL and the final unoccupied one satisfy the specific selection rule of $(\Delta n)_s = \pm 1$. This rule has been confirmed by magneto-optical measurements, as done for AB-stacked graphenes [18] and graphites [19, 20]. The perturbed LLs can present extra selection rules [3]. However, the undefined LLs show a large number of absorption peaks in the absence of a selection rule due to the random spatial distributions of the disordered modes. The above-mentioned magneto-excitation properties of LLs could be further examined by infrared transmission [18–20] and magneto-Raman spectroscopy [21].

The important differences between the generalized tight-binding model and the effective-mass approximation in magneto-electronic properties are worthy of detailed discussions. The latter might be suitable for some few-layer graphene with specific energy bands in the low-energy range. The effective-mass method is first to make the low-energy expansion at the K (K') point and then the magnetic quantization [5, 6, 22–29]. The effective Hamiltonian of monolayer graphene is

$$H(K, K') = \hbar v_F \begin{pmatrix} 0 & k_x \mp ik_y \\ k_x \pm ik_y & 0 \end{pmatrix}, \tag{3.3}$$

where $\mathbf{k} = (k_x, k_y)$ is measured from the K (K') point and $v_F = 3b\gamma_0/2$ indicates the fermi velocity at the conical points. The low-lying isotropic Dirac cones are characterized by $E^{c,v} = \pm v_F|\mathbf{k}|$, with an effective speed of $\sim C/300$ (C is light velocity). However, in few-layer graphene systems, one could derive the low-energy analytic dispersions when only considering the nearest-neighbor intralayer and interlayer atomic interactions (γ_0/β_0 and γ_1/β_1). In a similar way, the effective Hamiltonian of the AB-stacked bilayer graphene is expressed as

$$H(K, K') = \frac{\hbar^2}{2m^*} \begin{pmatrix} 0 & (k_x - ik_y)^2 \\ (k_x - ik_y)^2 & 0 \end{pmatrix}, \tag{3.4}$$

where the effective mass $m^* = |\gamma_1|/(3b\gamma_0/\sqrt{2})^2$. The parabolic energy dispersions $E^{c,v} = \pm\hbar^2 k^2/2m^*$ are straightforwardly obtained. With the increase of the layer number, electronic structures of AB-stacked graphenes could be regarded as the superposition of the monolayer and bilayer energy bands [30, 31]. As for ABC-stacked N-layer graphenes, the effective-mass model is only appropriate for the lowest pairs of conduction and valence bands crossing at the Fermi level, in which they behave as $E^{c,v} \propto \pm \gamma_0^N |\mathbf{k}|^N / |\gamma_1|^{N-1}$ [30, 31].

In the presence of a magnetic field, the effective Hamiltonian is derived by replacing $\mathbf{p}' = \mathbf{p} - e\mathbf{A}/c$ in the zero-field Hamiltonian. The calculated LL energies of monolayer, AB-stacked bilayer and ABC-stacked graphenes, respectively, display $\sqrt{B_z}, B_z$, and $\sqrt{B_z^N}$ dependences. It should be noted that the non-vertical interlayer atomic interactions would induce an infinite order of the effective Hamiltonian matrix for both AB- and ABC-stacked graphenes. Nevertheless, the effects beyond the minimal model could be realized through a qualitative perturbation analysis of these interlayer interactions [32, 33]. This method well describes the low-energy electronic/magneto-electronic properties related to the K and K' valleys. It is also suitable for studying low-energy LLs in monolayer silicene, germanene [34], MoS_2 [35–37], and few-layer phosphorene [38, 39]. However, a comprehensive description of the full energy spectrum could be achieved in the framework of the generalized Peierls tight-binding model, which simultaneously retains both the effects of external fields and all the important atomic interactions.

3.2 Experimental measurements

The theoretical calculations based on the generalized tight-binding model are compared with the up-to-date experimental measurements mainly being focused on angle-resolved photoemission spectroscopy (ARPES), [40–46] STM, [47–57] and optical spectroscopy [58–64, 64–71]. Experimental measurements can directly identify the main characteristics of the essential properties, providing a clear physical picture to establish the basic concepts. Furthermore, the complete and reliable results are very useful in the development of potential applications, such as electronic and optical devices [85, 86]. Provided with non-destructive spectroscopic techniques, these main instruments have been widely and appropriately employed for all the 2D materials studied throughout this book. They are very powerful in measuring the energy dispersions of valence bands [40–46], bond length [51, 57], corrugation [48, 53], lattice symmetry [47, 45], local nanostructure [49, 50], stacking configuration [47, 56], spin-split states [52, 54], spin polarization [52, 54], van Hove singularities in DOS [58, 60, 69–71] and the energy gap [69, 71, 81]. A brief introduction of the apparatus and recent experimental progress is presented.

ARPES has been widely used for directly measuring the wave-vector-dependent energy dispersions of solid-state materials [40–46]. The in-plane energy dispersions are derived straightforwardly in the principle of the photoelectric effect where the energy and parallel momentum are conserved, while the perpendicular component momentum is not conserved due to the breaking of translation symmetry along the normal direction. Accordingly, ARPES is mainly focused on 2D or quasi-2D materials with negligible normal dispersion relations, such as low-dimensional graphene-related systems, including graphene nanoribbons [46] and number- and stacking-dependent graphenes [40, 41, 43, 45]. Nevertheless, by interpreting the energy band structure in terms of the characteristics of the in-plane and out-plane dispersion relations, one can overcome the issue of the non-conserved normal momentum and obtain the 3D band structures of bulk materials, e.g. Bernal graphite [42, 44]. There are several verified spectroscopic characteristics including the energy gap and the 1D parabolic bands induced by quantum confinement in finite-width nanoribbons, [46] the Dirac-cone structure in monolayer graphene, [43] two/three pairs of linear bands in bilayer/tri-layer AA stacking, [40, 41] two pairs of parabolic bands in bilayer AB stacking, [43] the partially flat, sombrero-shaped and linear bands in tri-layer ABC stacking, [45] and the bilayer- and monolayer-like energy dispersions in Bernal graphite at the K and H points [42, 44]. The ARPES techniques are adopted for the other 2D materials studied in this book, and also can be used for simple hexagonal and rhombohedral graphites that exhibit the unsolved low-energy Dirac cones straightforwardly and spirally distributed along the K-H direction, respectively [87, 88]. Such experiments are helpful in determining the intralayer and interlayer hopping integrals and their significant effects.

Based on the quantum tunneling effect, STM was first invented by Binnig and Rohrer in 1982 [57]. Since then, it has become the most important experimental instrument in resolving the surface structure. With the magnitude exponentially decayed with the tip-surface distance, the tunneling current can be interpreted as the topographies of surfaces in real space with their lateral and vertical resolutions up to the atomic scale ($\sim$0.1 $\overset{\circ}{A}$) under an ultra-high vacuum environment [47, 55, 56]. From the examined materials the following are derived: detailed information of bond length [51, 57], corrugation [48, 53], lattice symmetry [47, 55], local nanostructure [49, 50], stacking configuration [47, 56] and spin-split states [52, 54]. Full spectroscopic information of the surface morphology has been obtained for many graphene-related materials, such as 0D fullerene, [53] 1D carbon nanotubes [48] and graphene nano-ribbons, [69–71] 2D graphenes [47, 56] and 3D graphites [51]. At the lateral and vertical directions, atomic resolution is useful in identifying different types in the various systems. It should be noted that STM is very sensitive to electron spin when probing a preselected domain with a ferromagnetic/antiferromagnetic tip. This spin-polarized STM was first proposed by Pierce in 1988 and realized by Wiesendanger et al in 1990 [52, 54]. Other experimental techniques are also available for characterizing the geometric properties of low-dimensional systems, such as the scanning transmission electron microscope (STEM), [89] transmission electron microscopy (TEM), [90] low-energy electron diffraction (LEED) [91] and atomic force microscopy (AFM) [92].

STS, an extension of STM, is used to demonstrate the tunneling current through the tip-surface junction in a constant height mode [93]. The electronic properties of the selected conducting surfaces are characterized based on the I–V and dI/dV curves sweeping over the bias voltage V. STS is an efficient method for examining the energy spectra of condensed-matter systems. The tunneling differential conductance (dI/dV) is approximately proportional to the DOS and directly presents the main features in DOS; useful information in identifying the intralayer and interlayer atomic interactions. Specifically, the low-lying DOS of monolayer graphene exhibits a linear ω-dependence with vanishing at the Dirac point [59]. As to the ABC-stacking configuration, a pronounced peak at the Fermi level is revealed in tri-layer and penta-layer cases as a result of the surface-localized states [60]. STS serves as a powerful experimental method for investigating the magnetically quantized energy spectra of layered graphenes. The measured tunneling differential conductance directly reflects the structure, energy, number and height of the LL peaks. Part of the theoretical predictions on the LL energy spectra are experimentally confirmed, such as the $\sqrt{B_z}$-dependent LL energy in monolayer graphene [58, 59, 61–63], the linear B_z-dependence in AB-stacked bilayer graphene [64–66], the coexistent square-root and linear B_z-dependences in tri-layer ABA stacking [64], and the 3D and 2D characteristics of the Landau subbands in AB-stacked graphite [67, 68].

Absorption [79, 80], transmission [72, 73, 79, 81] and reflection spectroscopies [74, 79] are the most versatile and commonly utilized spectroscopic techniques in the field of condensed-matter physics and material science. They are employed as analytical tools for a characterization of the optical properties of materials when the measurements are taken on the fraction of the incident radiation absorbed, transmitted, or reflected by a sample in a desired frequency range. A broadband light source is used and performed by a tungsten halogen lamp that has a broad range of modulation intensity and frequency [73, 74]. In magneto-optical experiments, the absorption intensity of monolayer graphene is proportional to the frequency as a result of the linear dispersion of the isotropic Dirac cone of massless Dirac fermions [72]. Besides, massive Dirac fermions are observed in AB-stacked bilayer graphene [79, 81]. Infrared reflection spectroscopy and absorption spectroscopy are also utilized to examine the partially flat and sombrero-shaped energy bands of ABC-stacked few-layer graphenes [80]. Moreover, the near-field optical microscopy combined with the scattering scan can distinguish the AB and ABC domains with a nano-scaled resolution due to their specific infrared conductivities [94].

Magnetic quantization phenomena of low-dimensional systems can also be studied by magneto-optical spectroscopies [72, 73, 82–84]. The magnetic field is performed by a superconducting magnet [72, 73, 84] and semidestructive single-turn coil [82, 83] with the desired field strength ranging from few T up to over one hundred T. The examined phenomena are exclusive in graphene-related systems, such as 0D LLs in few-layer graphenes [72, 73, 84] and 1D LLs in bulk graphites [72, 73, 84]. A lot of prominent delta-function-like absorption peaks are clearly revealed by the inter-LL excitations due to massless and massive Dirac fermions in monolayer [72] and AB-stacked bilayer graphenes [84]. The former and the latter absorption frequencies are square-root and linearly proportional to B_z, respectively. Concerning inter-Landau-subband excitations in Bernal graphite, one can observe a strong dependence on the wave vector k_z, which characterizes both kinds of Dirac quasi-particles [82, 83].

References

[1] Ho J H, Lai Y H, Chiu Y H and Lin M F 2008 Landau levels in graphene *Physica* E **40** 1722–5

[2] Lin C Y, Wu J Y, Ou Y J, Chiu Y H and Lin M F 2015 Magneto-electronic properties of multilayer graphenes *Phys. Chem. Chem. Phys.* **17** 26008–35

[3] Huang Y K, Chen S C, Ho Y H, Lin C Y and Lin M F 2014 Feature-rich magnetic quantization in sliding bilayer graphenes *Sci. Rep.* **4** 7509

[4] McClure J W 1956 Diamagnetism of graphite *Phys. Rev.* **104** 666

[5] Zheng Y and Ando T 2002 Hall conductivity of a two-dimensional graphite system *Phys. Rev.* B **65** 245420

[6] Gusynin V P and Sharapov S G 2005 Unconventional integer quantum Hall effect in graphene *Phys. Rev. Lett.* **95** 146801

[7] Do T N, Lin C Y, Lin Y P, Shih P H and Lin M F 2015 Configuration-enriched magnetoelectronic spectra of AAB-stacked trilayer graphene *Carbon* **94** 619–32

[8] Ho Y H, Tsai S J, Lin M F and Su W P 2013 Unusual Landau levels in biased bilayer bernal graphene *Phys. Rev.* B **87** 075417

[9] Lin Y P, Lin C Y, Chang C P and Lin M F 2015 Electric-field-induced rich magneto-absorption spectra of ABC-stacked trilayer graphene *RSC Adv* **5** 80410

[10] Lin C Y, Wu J Y, Chiu Y H and Lin M F 2014 Stacking-dependent magneto-electronic properties in multilayer graphenes *Phys. Rev.* B **90** 205434

[11] Lai Y H, Ho J H, Chang C P and Lin M F 2008 Magnetoelectronic properties of bilayer bernal graphene *Phys. Rev.* B **77** 085426

[12] Do T N, Lin C Y, Lin Y P, Shih P H and Lin M F 2015 Configuration-enriched magnetoelectronic spectra of AAB-stacked trilayer graphene *Carbon* **94** 619–32

[13] Inoue M 1962 Landau levels and cyclotron resonance in graphite *J. Phys. Soc. Japan* **17** 808

[14] Morimoto T and Koshino M 2013 Gate-induced dirac cones in multilayer graphenes *Phys. Rev.* B **87** 085424

[15] Xu P *et al* 2012 Electronic transition from graphite to graphene via controlled movement of the top layer with scanning tunneling microscopy *Phys. Rev.* B **86** 085428

[16] Xu P *et al* 2013 Graphene manipulation on 4H-SiC(0001) using scanning tunneling microscopy *Jpn. J. Appl. Phys.* **52** 035104

[17] Lin Y P, Lin C Y, Ho Y H, Do T N and Lin M F 2015 Magneto-optical properties of ABC-stacked trilayer graphene *Phys. Chem. Chem. Phys.* **17** 15921–7

[18] Orlita M *et al* 2008 Approaching the Dirac point in high-mobility multilayer epitaxial graphene *Phys. Rev. Lett.* **101** 267601

[19] Orlita M, Faugeras C, Schneider J M, Martinez G, Maude D K and Potemski M 2009 Graphite from the viewpoint of Landau level spectroscopy: an effective graphene bilayer and monolayer *Phys. Rev. Lett.* **102** 166401

[20] Orlita M, Faugeras C, Martinez G, Maude D K, Sadowski M L and Potemski M 2008 Dirac fermions at the H point of graphite: magnetotransmission studies *Phys. Rev. Lett.* **100** 136403

[21] Berciaud S, Potemski M and Faugeras C 2014 Probing electronic excitations in mono- to pentalayer graphene by micro magneto-raman spectroscopy *Nano Lett.* **14** 4548

[22] Sharapov S G, V P Gusynin V P and Beck H 2004 Magnetic oscillations in planar systems with the dirac-like spectrum of quasiparticle excitations *Phys. Rev.* B **69** 075104

[23] Goerbig M O 2011 Electronic properties of graphene in a strong magnetic field *Rev. Mod. Phys.* **83** 1193–243

[24] Chang C P 2011 Exact solution of the spectrum and magneto-optics of multilayer hexagonal graphene *J. Appl. Phys.* **110** 013725

[25] McCann E and Fal'ko V I 2006 Landau-level degeneracy and quantum Hall effect in a graphite bilayer *Phys. Rev. Lett.* **96** 086805

[26] Koshino M and McCann E 2011 Landau level spectra and the quantum Hall effect of multilayer graphene *Phys. Rev. B* **83** 165443

[27] H Min H and MacDonald A H 2008 Chiral decomposition in the electronic structure of graphene multilayers *Phys. Rev. B* **77** 155416

[28] Sena S H R, Pereira J M Jr, Peeters F M and Farias G A 2011 Landau levels in asymmetric graphene trilayers *Phys. Rev. B* **84** 205448

[29] McCann E and Koshino M 2013 The electronic properties of bilayer graphene *Rep Prog. Phys.* **76** 056503

[30] Min H and MacDonald A H 2008 Electronic structure of multilayer graphene *Prog. Theor. Phys. Supp* **176** 227

[31] Manes J L, Guinea F and Vozmediano M A H 2007 Existence and topological stability of fermi points in multilayered graphene *Phys. Rev. B* **75** 155424

[32] Koshino M and Ando T 2008 Magneto-optical properties of multilayer graphene *Phys. Rev. B* **77** 115313

[33] Koshino M and McCann E 2009 Trigonal warping and Berry's phase N^π in ABC-stacked multilayer graphene *Phys. Rev. B* **80** 165409

[34] Ezawa M 2012 Valley-polarized metals and quantum anomalous Hall effect in silicene *Phys. Rev. Lett.* **109** 055502

[35] Tahir M, Vasilopoulos P and Peeters F M 2016 Quantum magnetotransport properties of a MoS_2 monolayer *Phys. Rev. B* **93** 035406

[36] Kormányos A, Rakyta P and Burkard G 2015 Landau levels and Shubnikov-de Haas oscillations in monolayer transition metal dichalcogenide semiconductors *New J. Phys.* **17** 103006

[37] Li X, Zhang F and Niu Q 2013 Unconventional quantum Hall effect and tunable spin Hall effect in Dirac materials: application to an isolated MoS_2 trilayer *Phys. Rev. Lett.* **110** 066803

[38] Pereira J M Jr and Katsnelson M I 2015 Landau levels of single-layer and bilayer phosphorene *Phys. Rev. B* **92** 075437

[39] Jiang Y, Roldán R, Guinea F and Low T 2015 Magnetoelectronic properties of multilayer black phosphorus *Phys. Rev. B* **92** 085408

[40] Kim K S *et al* 2013 Coexisting massive and massless Dirac fermions in symmetry-broken bilayer graphene *Nat. Mater.* **12** 887

[41] Bao C, Yao W, Wang E, Chen C, Avila J, Asensio M C and Zhou S 2017 Stacking-dependent electronic structure of trilayer graphene resolved by nanospot angle-resolved photoemission spectroscopy *Nano Lett.* **8** 1564

[42] Gruneis A *et al* 2008 Electron–electron correlation in graphite: a combined angle-resolved photoemission and first-principles study *Phys. Rev. Lett.* **100** 037601

[43] Ohta T, Bostwick A, Seyller T, Horn K and Rotenberg E 2006 Controlling the electronic structure of bilayer graphene *Science* **313** 951

[44] Zhou S Y *et al* 2006 First direct observation of Dirac fermions in graphite *Nat. Phys.* **2** 595

[45] Coletti C *et al* 2013 Revealing the electronic band structure of trilayer graphene on SiC: an angle-resolved photoemission study *Phys. Rev. B* **88** 155439

[46] Ruffieux P, Cai J, Plumb N C, Patthey L, Prezzi D, Ferretti A and Fasel R 2012 Electronic structure of atomically precise graphene nanoribbons *ACS Nano* **6** 6930

[47] Que Y, Xiao W, Chen H, Wang D, Du S and Gao H J 2015 Stacking-dependent electronic property of trilayer graphene epitaxially grown on Ru(0001) *Appl. Phys. Lett.* **107** 263101

[48] Odom T W, Huang J-L, Kim P and Lieber C M 1998 Atomic structure and electronic properties of single-walled carbon nanotubes *Nature* **391** 62

[49] González-Herrero H *et al* 2016 Graphene tunable transparency to tunneling electrons: a direct tool to measure the local coupling *ACS Nano* **10** 5131

[50] Andryushechkin B V, Shevlyuga V M, Pavlova T V, Zhidomirov G M and Eltsov K N 2016 Adsorption of O_2 on Ag(111): evidence of local oxide formation *Phys. Rev. Lett.* **117** 056101

[51] Zha B *et al* 2017 Cooperation and competition between halogen bonding and van der Waals forces in supramolecular engineering at the aliphatic hydrocarbon/graphite interface: position and number of bromine group effects *Nanoscale* **9** 237

[52] Pierce D T 1988 Spin-polarized electron microscopy *Phys. Scr.* **38** 291

[53] Matsumoto M *et al* 2004 Adlayers of C60–C60 and C60–C70 fullerene dimers formed on Au (111) in benzene solutions studied by STM and LEED *Langmuir* **20** 1245

[54] Wiesendanger R, Güntherodt H-J, Güntherodt G, Gambino R and Ruf R 1990 Observation of vacuum tunneling of spin-polarized electrons with the scanning tunneling microscope *Phys. Rev. Lett.* **65** 247

[55] Xu R, Yin L J, Qiao J B, Bai K K, Nie J C and He L 2015 Direct probing of the stacking order and electronic spectrum of rhombohedral trilayer graphene with scanning tunneling microscopy *Phys. Rev.* B **91** 035410

[56] Mondelli P, Gupta B, Betti M G, Mariani C, Duffin J L and Motta N 2017 High quality epitaxial graphene by hydrogenetching of 3C-SiC(111) thin-film on Si(111) *Nanotechnology* **28** 115601

[57] Binnig G and Rohrer H 1986 Scanning tunneling microscopy *IBM J. Res. Dev.* **30** 355

[58] Miller D L, Kubista K D, Rutter G M, Ruan M, de Heer W A, First P N and Stroscio J A 2009 Observing the quantization of zero mass carriers in graphene *Science* **324** 924

[59] Li G H, Luican A and Andrei E Y 2009 Scanning tunneling spectroscopy of graphene on graphite *Phys. Rev. Lett.* **102** 176804

[60] Pierucci D *et al* 2015 Evidence for flat bands near the Fermi level in epitaxial rhombohedral multilayer graphene *ACS Nano* **9** 5432

[61] Luican A, Li G H, Reina A, Kong J, Nair R R, Novoselov K S, Geim A K and Andrei E Y 2011 Single-layer behavior and its breakdown in twisted graphene layers *Phys. Rev. Lett.* **106** 126802

[62] Song Y J *et al* 2010 High-resolution tunnelling spectroscopy of a graphene quartet *Nature* **467** 185

[63] Wang W X, Yin L J, Qiao J B, Cai T, Li S Y, Dou R F, Nie J C, Wu X S and He L 2015 Atomic resolution imaging of the two-component Dirac–Landau levels in a gapped graphene monolayer *Phys. Rev.* B **92** 165420

[64] Yin L J, Li S Y, Qiao J B, Nie J C and He L 2015 Landau quantization in graphene monolayer, bernal bilayer, and bernal trilayer on graphite surface *Phys. Rev.* B **91** 115405

[65] G M Rutter G M, S. Jung S, Klimov N N, Newell D B, Zhitenev N B and Stroscio J A 2011 Microscopic polarization in bilayer graphene *Nat. Phys.* **7** 649

[66] Yin L J, Zhang Y, Qiao J B, Li S Y and He L 2016 Experimental observation of surface states and Landau levels bending in bilayer graphene *Phys. Rev.* B **93** 125422

[67] T Matsui T, Kambara H, Niimi Y, Tagami K, Tsukada M and Fukuyama H 2005 STS observations of Landau levels at graphite surfaces *Phys. Rev. Lett.* **94** 226403

[68] Li G H and Andrei E Y 2007 Observation of Landau levels of dirac fermions in graphite *Nat. Phys.* **3** 623

[69] Sugiyama Y *et al* 2014 Spectroscopic study of graphene nanoribbons formed by crystallographic etching of highly oriented pyrolytic graphite *Appl. Phys. Lett.* **105** 123116

[70] Kravets V G *et al* 2010 Spectroscopic ellipsometry of graphene and an exciton-shifted van hove peak in absorption *Phys. Rev.* B **81** 155413

[71] Miccoli I *et al* 2017 Quasi-free-standing bilayer graphene nanoribbons probed by electronic transport *Appl. Phys. Lett.* **110** 051601

[72] Jiang Z *et al* 2007 Infrared spectroscopy of Landau levels of graphene *Phys. Rev. Lett.* **98** 197403

[73] Plochocka P *et al* 2008 High-energy limit of massless dirac fermions in multilayer graphene using magneto-optical transmission spectroscopy *Phys. Rev. Lett.* **100** 087401

[74] Mak K F, Sfeir M Y, Wu Y, Hung Lui C, Misewich J A and Heinz T F 2008 Measurement of the optical conductivity of graphene *Phys. Rev. Lett.* **101** 196405

[75] Li Z Q *et al* 2009 Band structure asymmetry of bilayer graphene revealed by infrared spectroscopy *Phys. Rev. Lett.* **102** 037403

[76] Mak K F, Shan J and Heinz T F 2010 Electronic structure of few-layer graphene: experimental demonstration of strong dependence on stacking sequence *Phys. Rev. Lett.* **104** 176404

[77] Mak K F, Lui C H, Shan J and Heinz T F 2009 Observation of an electric-field-induced band gap in bilayer graphene by infrared spectroscopy *Phys. Rev. Lett.* **102** 256405

[78] Plochocka P, Solane P Y, Nicholas R J, Schneider J M, Piot B A, Maude D K, Portugall O and Rikken G L J A 2012 Origin of electron-hole asymmetry in graphite and graphene *Phys. Rev.* B **85** 245410

[79] Li Z Q *et al* 2009 Band structure asymmetry of bilayer graphene revealed by infrared spectroscopy *Phys. Rev. Lett.* **102** 037403

[80] Mak K F, Shan J and Heinz T F 2010 Electronic structure of few-layer graphene: experimental demonstration of strong dependence on stacking sequence *Phys. Rev. Lett.* **104** 176404

[81] Mak K F, Lui C H, Shan J and Heinz T F 2009 Observation of an electric-field-induced band gap in bilayer graphene by infrared spectroscopy *Phys. Rev. Lett.* **102** 256405

[82] Plochocka P, Solane P Y, Nicholas R J, Schneider J M, Piot B A, Maude D K, Portugall O and Rikken G L J A 2012 Origin of electron-hole asymmetry in graphite and graphene *Phys. Rev.* B **85** 245410

[83] Nicholas R J, Solane P Y and Portugall O 2013 Ultrahigh magnetic field study of layer split bands in graphite *Phys. Rev. Lett.* **111** 096802

[84] Orlita M *et al* 2011 Magneto-optics of bilayer inclusions in multilayered epitaxial graphene on the carbon face of SiC *Phys. Rev.* B **83** 125302

[85] Szafranek B N, Fiori G, Schall D, Neumaier D and Kurz H 2012 Current saturation and voltage gain in bilayer graphene field effect transistors *Nano Lett.* **12** 1324

[86] Liu M *et al* 2011 A graphene-based broadband optical modulator *Nature* **474** 64

[87] Ho C H, Chang C P, Su W P and Lin M F 2013 Precessing anisotropic Dirac cone and Landau subbands along a nodal spiral *New J. Phys.* **15** 053032

[88] Chen R B and Chiu Y H 2013 Landau subband and Landau level properties of AA-stacked graphene superlattice *J. Nanosci. Nanotechnol.* **12** 2557–66

[89] Liu Z B *et al* 2017 Phase transition and *in situ* construction of lateral heterostructure of 2D superconducting α/β Mo$_2$C with sharp interface by electron beam irradiation *Nanoscale* **9** 7501

[90] Lin Q Y *et al* 2012 High-resolution TEM observations of isolated rhombohedral crystallites in graphite blocks *Carbon* **50** 2369

[91] Dai Z W *et al* 2017 Surface structure of bulk 2H-MoS$_2$ (0001) and exfoliated suspended monolayer MoS$_2$: a selected area low energy electron diffraction study *Surf. Sci.* **660** 16

[92] Miyazawa K, Watkins M, Shluger A L and Fukuma T 2017 Influence of ions on two-dimensional and three-dimensional atomic force microscopy at fluorite-water interfaces *Nanotechnology* **28** 245701

[93] Chen C J 1993 *Introduction to Scanning Tunneling Microscopy* (Oxford: Oxford University Press)

[94] Kim D S *et al* 2015 Stacking structures of few-layer graphene revealed by phase-sensitive infrared nanoscopy *ACS Nano* **9** 6765

IOP Publishing

Theory of Magnetoelectric Properties of 2D Systems

S C Chen, J Y Wu, C Y Lin and M F Lin

Chapter 4

Silicene, germanene and tinene

For monolayer silicene, germanene and tinene, their Hamiltonians cover the interactions related to the sp^3 bondings, the SOCs and the buckled structures, being absent in monolayer graphene. They can induce unusual band structures and the novel quantization phenomena, especially for tinene with the strongest orbital hybridizations and SOCs. Such systems present the SOC-induced separation of Dirac points at the K/K' valley. In particular, the combined effects in tinene lead to the low-lying valence and conduction bands near the Γ point. Only tinene presents two groups of LLs near the Fermi level, respectively corresponding to the p$_z$- and (p$_x$,p$_y$)-dominated electronic states. There are certain important differences between them in terms of orbital components, spin configurations, localization centers, state degeneracy, and magnetic- and electric-field dependencies. The LL splittings in the first and second groups mainly arise from the effects of electric and magnetic fields, respectively. Specifically, the LL anti-crossings only appear in the first group during a variation of the electric field, in which the critical mechanism, the relationship between the electric field and SOC, is responsible for the probability distribution transfers among the orbital components/distinct sublattices/spin configurations. The E_z-induced LL splittings are also present in silicene and germanene. However, the anti-crossing behaviors are only revealed in the latter.

For the group-IV inorganic layered systems, the sp^3 orbital bondings and the SOCs are included in the critical Hamiltonians. In the bases of $\{|p_z^A\rangle, |p_x^A\rangle, |p_y^A\rangle, |s^A\rangle, |p_z^B\rangle, |p_x^B\rangle, |p_y^B\rangle, |s^B\rangle\} \otimes \{\uparrow, \downarrow\}$, the nearest-neighbor Hamiltonian of the monolayer system is expressed as

$$H = \sum_{\langle I\rangle, o, m} E_o C_{Iom}^{+} C_{Iom} + \sum_{\langle I,J\rangle, o,o', m} \gamma_{oo'}^{\Delta\mathbf{R}_I} C_{Iom}^{+} C_{Jo'm}$$
$$+ \sum_{\langle I\rangle, p_\alpha, p_\beta, m, m'} \frac{\lambda_{\text{SOC}}}{2} C_{Ip_\alpha m}^{+} C_{Ip_\beta m'} (-i\varepsilon_{\alpha\beta\gamma}\sigma_{mm'}^{\gamma}). \tag{4.1}$$

The first, second and third terms are, respectively, the site energy, the nearest-neighbor hopping integral and SOC. The site energy and the nearest-neighbor hopping integrals for Si (Ge) are $E_{3s} = -7.03$ eV ($E_{4s} = -8.02$ eV), $V_{ss\sigma} = -1.93$ eV (-1.79 eV), $V_{sp\sigma} = 2.54$ eV (2.36 eV), $V_{pp\sigma} = 4.47$ eV (4.15 eV), and $V_{pp\pi} = -1.12$ eV (-1.04 eV) [1]. The SOC strength is, respectively, predicted to be $\lambda_{SOC} = 0.034$ and 0.196 eV's for Si and Ge, respectively.

The SOC, buckled structure, and orbital hybridizations in group-IV layered systems can induce feature-rich energy bands and diversify the quantized LLs. Germanene and silicene have similar band structures (black and green curves in figure 4.1(a)), in which the low-lying electronic states mainly come from the $4p_z$ and $3p_z$ orbitals, respectively. A small direct energy gap, which corresponds to the slightly separated Dirac points, is dependent on the strength of SOC. E_g is, respectively, 45 meV and 5 meV for Ge and Si systems (inset in figure 4.1(a)). The first pair of valence and conduction bands have doubly degenerate states associated with the spin-down- and spin-up-dominated equivalent configurations. It is sufficient to only discuss one of these two configurations, as shown in figures 4.1(b) and (c) for germanene. Near the K (K') point, the valence states are mainly determined by the $|4p_z^B; \downarrow\rangle$ and $|4p_z^A; \downarrow\rangle$ ($|4p_z^B; \uparrow\rangle$ and $|4p_z^A; \uparrow\rangle$). Their contributions are very sensitive to the changes of wave vectors along K$\rightarrow$M(K'$\rightarrow\Gamma$)) (black curves in figure 4.1(b)). A similar behavior is revealed in the conduction states under the interchange of the A and B sublattices (figure 4.1(c)). In addition, the ($4p_x$,$4p_y$,4s) orbitals can make important contributions to the middle-energy states close to the Γ point.

The quantized LLs in monolayer germanene (silicene) are characterized by the subenvelope functions on the A and B sublattices with sp^3 orbitals and two spin configurations. All the low-lying LLs in monolayer germanene (silicene) belong to the well-behaved modes. They are eight-fold degenerate for each (k_x,k_y) state, except the four-fold degenerate LLs of $n^{c,v} = 0$. As to each localization center, there are two subgroups characterized by the up- and down-dominated configurations, as indicated in figures 4.2(a)–(d) for the 2/6 states. The first and the second subgroups, respectively, have the $n^c = 0$ conduction LL and the $n^v = 0$ valence LL at $E^c = 23$ meV and $E^v = -23$ meV (figure 4.2(a)). The former and the latter are caused by the spin-up and spin-down configurations in the dominating B sublattice, respectively (figure 4.2(b) and (d)). The other $n^{c,v} \neq 0$ LLs in these two subgroups are doubly degenerate, and their wave functions are identical under the interchanges of spins and weights of the A and B sublattices. Certain important differences exist between germanene and graphene (figures 4.2(a)–(d) and figures 3.1(c) and (d)). Germanene exhibits significantly split $n^{c,v}$=0 LLs with partial contributions from the A sublattice. The weight ratio between the A and B sublattices is quite different for the valence and conduction LLs. In addition to the dominating $4p_z$ orbitals, the contributions due to the ($4p_x$,$4p_y$,4s) orbitals are gradually enhanced as $|E^{c,v}\rangle$ grows. However, the opposite is true for graphene.

A perpendicular electric field applied to buckled systems can split energy bands and even induce anti-crossings in the LL spectra. In addition, band structures of

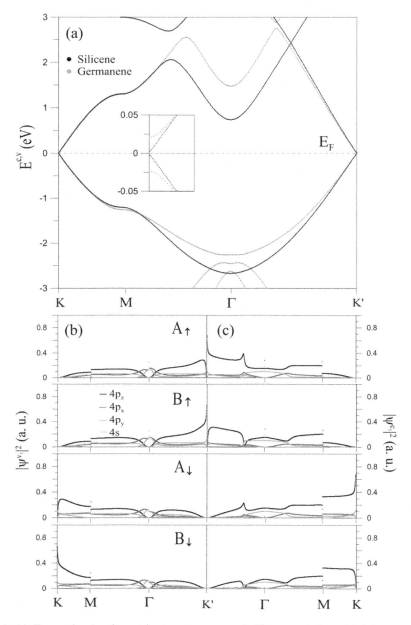

Figure 4.1. (a) Energy bands of monolayer germanene and silicene, and the orbital-decomposed state probabilities along the high-symmetry points for the first pair of (b) valence and (c) conduction bands. The inset of (a) is the band structure near the Dirac point.

layered graphenes do not present the splitting behavior under E_z, mainly owing to the absence of SOC. The destruction of the $z = 0$ mirror symmetry causes one Dirac cone to split into two cones (black curves in figure 4.3(a)) when the gate voltage between two sublattices grows from zero. The lower cone structure approaches the

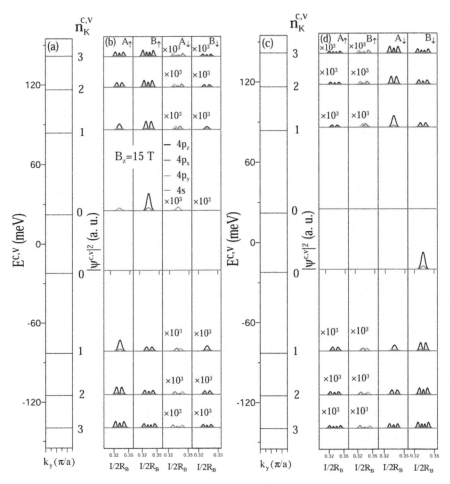

Figure 4.2. For germanene at $B_z = 15$ T, (a) and (b) the up-dominated and (c) and (d) the down-dominated LL energies and spatial probability distributions, corresponding to the quantized K-valley states.

Fermi level, and the energy gap vanishes at a critical V_c where the linearly gapless Dirac-cone structure is recovered (blue curves in figure 4.3(a)). The dependence of E_g on V_z is in the cusp form (figure 4.3(b)). Another cone structure lies always away from E_F. The V_z-dependent cone structures are quantized into unusual LL energy spectra (figure 4.3(c)). The K-valley-dependent (or the K'-valley-dependent) LLs are split according to the magnetic quantization of the lower and higher Dirac-cone structures. The $n^{c,v} > 0$ and $n^{c,v} = 0$ LLs, respectively, have four-fold and double degeneracy. The split LL energy spectrum, which corresponds to the lower Dirac cone, exhibits a non-monotonous V_z-dependence. As a result, intragroup LL anti-crossings occur frequently in the plentiful LL energy spectrum (large circles). In addition, the V_z-induced LL splittings and anti-crossings are also presented in layered graphenes, except for the AA-stacked systems [2].

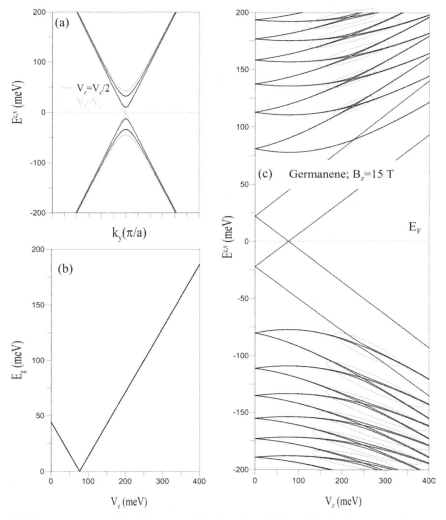

Figure 4.3. For germanene, (a) low-lying energy bands at $V_z = V_c/2$ and V_c, (b) the V_z-dependence of energy gap, and (c) the V_z-dependent LL energy spectra at $B_z = 15$ T. The LL anti-crossings are indicated by the red circles.

The close cooperation of the electric field and spin-orbital interactions can create a significant probability transfer between the spin-up and spin-down configurations and thus frequent intragroup LL anti-crossings. For example, the $n_K^v = 2$ and $n_K^v = 3$ LLs exhibit dramatic changes in the spatial distributions within the critical range of 130 meV $< V_z <$ 370 meV (green and purple triangles in figure 4.4(b)). At small V_z's, these two $4p_z$-dominated LLs have similar quantum modes on the $(A_\uparrow, B_\uparrow)$ and $(A_\downarrow, B_\downarrow)$ sublattices (figures 4.4(a) and (c)). However, the weight of distinct spin configurations is very large and small for the former and the latter, respectively. With the variation of V_z, the electric field can induce the probability transfer between the $A_\uparrow$ and $B_\uparrow$ ($A_\downarrow$ and $B_\downarrow$) sublattices. Furthermore, the intra-atomic SOC

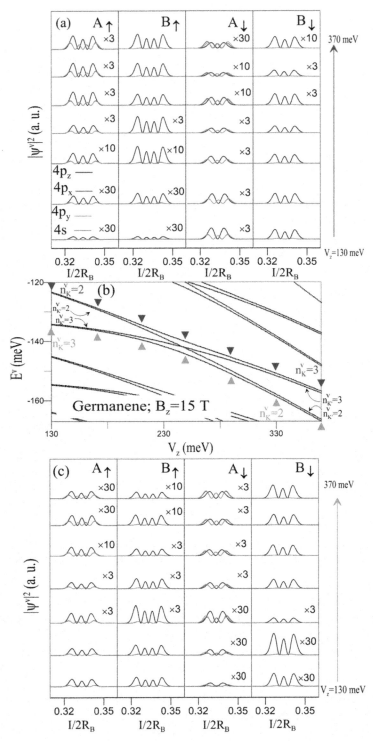

Figure 4.4. For germanene, the LL crossings and anti-crossings (b) within a certain range of E^v, accompanied with (a) and (c) the drastic changes of probability distributions during the LL anti-crossings.

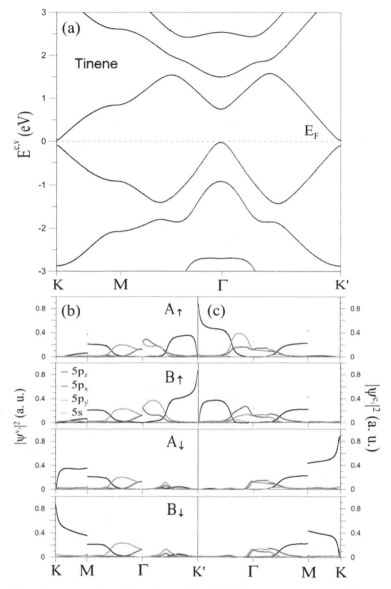

Figure 4.5. (a) Band structure of monolayer tinene, the orbital-decomposed state probabilities along the high symmetry points for the first pair of (b) valence and (c) conduction bands.

of $4p_z$ and $(4p_x, 4p_y)$ orbitals results in the significant distribution change on the $A_\uparrow$ and $A_\downarrow$ ($B_\uparrow$ and $B_\downarrow$) sublattices. This means that the latter two orbitals play a critical role in the anti-crossing behavior, even if they have small weight. The comparable probability distributions on the spin-related sublattices are the main reason for the anti-crossings of the n_K^v and n_K^v+1 LLs (the $n_{K'}^c$ and $n_{K'}^c+1$ LLs). However, the direct crossing from the $n_{K'}^v$ and $n_{K'}^v+1$ LLs (the n_K^c and n_K^c+1 LLs) occurs simultaneously.

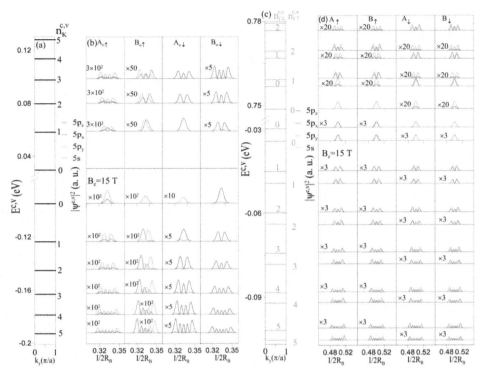

Figure 4.6. For tinene, the K-dependent LL (a) energies and (b) probability distributions, and same plots in (c) and (d) for the Γ-dependent LLs. The K-dependent LLs are spin degenerate except for $n_K^{c,v} = 0$. In (b) the down-dominated LL spatial probability distributions are shown.

Tinene has more low-lying energy bands and diverse LLs, compared with germanene and silicene. A pair of slightly distorted Dirac cones appears near the K point (figure 4.5(a)), as observed in germanene and silicene (figure 4.1(a)). Low-lying parabolic bands initiated at the Γ point also exist, being absent in the other two systems. Such energy bands, which mainly originate from the $(5p_x, 5p_y)$ orbitals (figures 4.5(b) and (c)), are induced by the stronger sp^3 bonding. State degeneracy at the Γ point is further destroyed by the critical SOC between the $5p_x$ and $5p_y$ orbitals [3], so that one of the parabolic bands is very close to the Fermi level. The Dirac-cone structure is quantized into the first group of LLs (the black lines in figure 4.6(a)), in which the main features are similar to those in germanene (figures 4.2(b) and (d)), such as, the p_z-orbital dominance (the black curves in figure 4.6(b)), localization centers, state degeneracy, spin configurations, quantum modes on the A and B sublattices, and B_z-dependence of LL energy spectrum. On the other side, the first group is in sharp contrast to the second group (the blue lines in figure 4.6(c)). Both $5p_x$ and $5p_y$ orbitals dominate the second group of LLs and make almost the same contributions (red and green curves in figure 4.6(d)). Such LLs only have two equivalent centers of 1 and 1/2, and they are doubly degenerate. For each center, the split LLs are characterized by the up- and down-dominated configurations ($n_{\Gamma,\uparrow}^{c,v}$ and

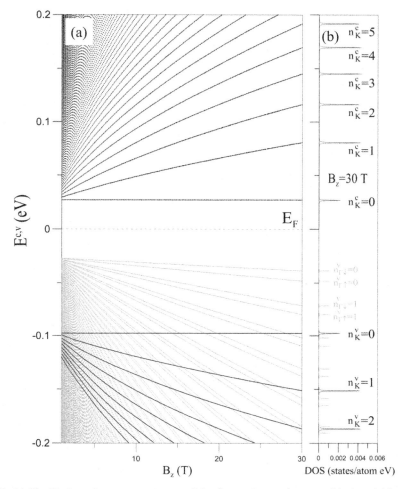

Figure 4.7. (a) The B_z-dependent energy spectra of the first and second groups (black and blue curves) in tinene, and (b) density of states at $B_z = 30$ T.

$n_{\Gamma,\downarrow}^{c,v}$), being attributed to the significant effect of SOC. The spin-split LL energy spacing is observable, especially for the larger spacing in the conduction LLs. This spacing grows with an increased weight ratio of two spin configurations on the same sublattice. Specifically, the A and B sublattices present the same quantum modes, since the nearest-neighbor hopping integrals near the Γ point are roughly proportional to the square of wave vector. However, the hexagonal symmetry can generate the linear k-dependence in these atomic interactions near the K point. As to the first group, this accounts for the mode difference of one between the A and B sublattices.

Tinene exhibits a rich B_z-dependent energy spectrum and density of states. Two groups of LLs have well-behaved modes, as indicated by the absence of anti-crossings and the existence of intergroup crossings (figure 4.7(a)). The LL state energies grow with an increase of field strength except for those almost unchanged $n_K^{c,v} = 0$. As to the first and the second groups, the B_z-dependence is presented in the

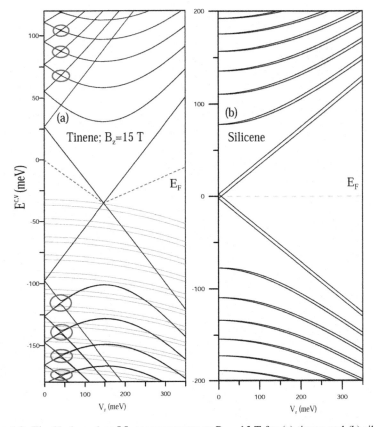

Figure 4.8. The V_z-dependent LL energy spectra at $B_z = 15$ T for (a) tinene and (b) silicene.

square-root and the linear forms, respectively (black and blue curves). This directly reflects the magnetic quantization from the linear and the parabolic energy dispersions. The spin-split energy spacings in the second group gradually become large, since the higher field strength creates more localized LL wave functions and enhances the spin-up or spin-down dominance. That is to say, the splitting energies are enlarged by the stronger effects of the SOC. The main differences between the two groups of energy spectra are further revealed in the DOS (figure 4.7(b)). A lot of strong peaks appear in the delta-function-like symmetric structure, reflecting the discrete characteristics of LLs. Their heights are proportional to the state degeneracy. The single- and double-peak structures, respectively, originate from the first and the second groups of LLs (black and blue curves); furthermore, the former have larger peak spacings. The main features of LL DOS, peak structure, height, number and energy, could be verified from STS measurements [4–8].

The V_z-dependent LL energy spectra are quite different among the group-IV layered systems. The split LLs cannot survive only in the AA-stacked graphenes, since the mirror symmetry is preserved even in the composite magnetic and electric fields [2]. For monolayer silicene, the splitting energy spacings are very small, and the

LL anti-crossings and crossings are absent (figure 4.8(b)). The weak SOC and the large v_F (the strong energy dispersion) are responsible for the monotonous V_z-dependence. However, monolayer germanene and tinene frequently exhibit intragroup anti-crossings and crossings (red circles in figures 4.3(c) and 4.8(a)), in which two anti-crossing LLs have a quantum number difference of $\Delta n = 1$. Specifically, the latter have intergroup crossings between the $5p_z$- and ($5p_x$,$5p_y$)-dominated LLs (black and blue curves in figure 4.8(a)). The V_z-induced intragroup anti-crossings are also observed in non-AA-stacked graphenes, while they arise from two LLs with $\Delta n = 3i$ (figure 3.7(b)) [9–12]. In addition to V_z, $\Delta n = 1$ and $3i$ are, respectively, determined by the significant spin-orbital interactions and certain interlayer hopping integrals.

The electronic properties of the group-IV layered systems may be validated by the measurements from ARPES [13, 14]. The anti-crossing and crossing LL behaviors would be reflected in many other physical properties, such as the optical absorption spectrum and the plasmon excitations, which may be obtained by performing the EELS reflection [15, 16], as well as the inelastic light scattering [17, 18]. Moreover, the V_z-induced LL splittings could create extra steps in the magnetotransport measurement.

References

[1] Liu C C, Jiang H and Yao Y 2011 Low-energy effective Hamiltonian involving spin-orbit coupling in silicene and two-dimensional germanium and tin *Phys. Rev.* B **84** 195430

[2] Tsai S J, Chiu Y H, Ho Y H and Lin M F 2012 Gate-voltage-dependent Landau levels in AA-stacked bilayer graphene *Chem. Phys. Lett.* **550** 104–10

[3] Chen S C, Wu C L, Wu J Y and Lin M F 2016 Magnetic quantization of sp³ bonding in monolayer gray tin *Phys. Rev.* B **94** 045410

[4] Li G H, Luican A and Andrei E Y 2009 Scanning tunneling spectroscopy of graphene on graphite *Phys. Rev. Lett.* **102** 176804

[5] Miller D L, Kubista K D, Rutter G M, Ruan M, de Heer W A, First P N and Stroscio J A 2009 Observing the quantization of zero mass carriers in graphene *Science* **324** 924

[6] Luican A, Li G H, Reina A, Kong J, Nair R R, Novoselov K S, Geim A K and Andrei E Y 2011 Single-layer behavior and its breakdown in twisted graphene layers *Phys. Rev. Lett.* **106** 126802

[7] Song Y J *et al* 2010 High-resolution tunnelling spectroscopy of a graphene quartet *Nature* **467** 185

[8] Wang W X, Yin L J, Qiao J B, Cai T, Li S Y, Dou R F, Nie J C, Wu X S and He L 2015 Atomic resolution imaging of the two-component Dirac–Landau levels in a gapped graphene monolayer *Phys. Rev.* B **92** 165420

[9] Lin C Y, Wu J Y, Chiu Y H and Lin M F 2014 Stacking-dependent magneto-electronic properties in multilayer graphenes *Phys. Rev.* B **90** 205434

[10] Lin C Y, Wu J Y, Ou Y J, Chiu Y H and Lin M F 2015 Magneto-electronic properties of multilayer graphenes *Phys. Chem. Chem. Phys.* **17** 26008–35

[11] Inoue M 1962 Landau levels and cyclotron resonance in graphite *J. Phys. Soc. Japan* **17** 808

[12] Morimoto T and Koshino M 2013 Gate-induced Dirac cones in multilayer graphenes *Phys. Rev.* B **87** 085424

[13] Zhu F, Chen W, Xu Y, Gao C, Guan D, Liu C, Qian D, Zhang S C and Jia J 2015 Epitaxial growth of two-dimensional stanene *Nat. Mater.* **14** 1020–5

[14] Mahatha S K, Moras P, Bellini V, Sheverdyaeva P M, Struzzi C, Petaccia L and Carbone C 2014 Silicene on Ag(111): a honeycomb lattice without Dirac bands *Phys. Rev.* B **89** 201416

[15] Wachsmuth P, Hambach R, Kinyanjui M K, Guzzo M, Benner G and Kaiser U 2013 High-energy collective electronic excitations in free-standing single-layer graphene *Phys. Rev.* B **88** 075433

[16] Pan C T, Nair R R, Bangert U, Ramasse Q, Jalil R, Zan R, Seabourne C R and Scott A J 2012 Nanoscale electron diffraction and plasmon spectroscopy of single- and few-layer boron nitride *Phys. Rev.* B **85** 045440

[17] Devereaux T and Hackl R 2007 Inelastic light scattering from correlated electrons *Rev. Mon. Phys.* **79** 175

[18] García-Flores A F, Terashita H, E Granado E and Kopelevich Y 2009 Landau levels in bulk graphite by Raman spectroscopy *Phys. Rev.* B **79** 113105

Chapter 5

Few-layer phosphorenes

A phosphorene layer is a puckered structure. The deformed hexagonal lattice in the x–y plane is quite different from the honeycomb lattice of group-IV systems. This unique geometric structure fully dominates the low-lying energy bands which are highly anisotropic in the energy dispersion and magnetic quantization. A perpendicular electric field can significantly diversify the electronic properties. The electric and magnetic fields can create diverse phenomena in a bilayer system, such as two subgroups of Landau levels (LLs), uniform and non-uniform LL energy spacings, and frequent crossings and anti-crossings. These novel results are clearly explained by the spatial distributions of the subenvelope functions.

The band structures of few-layer phosphorene near the Fermi level are mainly determined by the $3p_z$-orbital hybridizations [1]. Their Hamiltonians, as discussed in section 2.4, are given by

$$H = \sum_{\langle IJ \rangle \langle ll' \rangle} - h_{IJ}^{ll'} C_{Il}^{+} C_{Jl'},$$

(5.1)

where $h_{IJ}^{ll'}$ represents the five intralayer (figure 2.2(a)) and four interlayer hopping integrals (figures 2.3(a) and (b); details in [1]). The cooperation of the complicated multi-hopping integrals and the external fields can greatly diversify electronic structures and LLs.

Monolayer phosphorene has a direct gap of ~1.6 eV near the Γ point (figure 5.1(a)), being in sharp contrast with that dominated by the K point in the group-IV systems (figures 3.1(a) and 4.1(a)). The first pair of energy bands nearest to E_F is, respectively, linear and parabolic along the ΓX and ΓY directions. The valence (conduction) band is due to the linearly anti-symmetric (symmetric) superposition of the tight-binding functions on the upper and lower subplanes. As to bilayer phosphorene, two pairs of low-lying bands have parabolic dispersions, as shown in figure 5.1(b).

doi:10.1088/978-0-7503-1674-3ch5

© IOP Publishing Ltd 2017

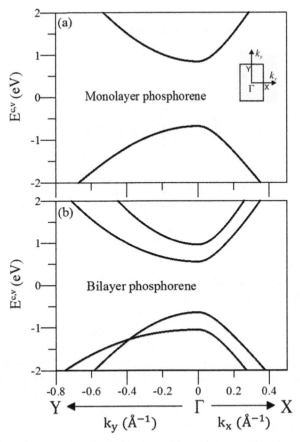

Figure 5.1. Energy bands of (a) monolayer and (b) bilayer phosphorene. The first Brillouin zone is also shown in (a).

The quantized LLs in phosphorene are characterized by the subenvelope functions on the different subplanes and layers. They are localized at the 1/2 and 2/2 positions of the enlarged unit cell, corresponding to the magnetic quantization initiated from the Γ point. All the sublattices present similar oscillation modes with comparable probability distributions (or amplitudes). The well-behaved spatial distributions, as shown in figures 5.2(b) and (c), are similar to those of monolayer graphene (figure 3.1(d)). The $3p_z$-orbital quantization, with spin degree, is four-fold degenerate for each (k_x, k_y) state. This is in great contrast to the eight-fold degeneracy in the group-IV systems (chapters 3 and 4), or the double degeneracy of the spin- and valley-dependent LLs in MoS_2 (chapter 6). The LL degeneracy depends on the number of equivalent valleys, the existence of inversion symmetry ($z \rightarrow -z$ and $x \rightarrow -x$), and spin configuration. There are two groups of valence and conduction LLs in bilayer phosphorene (black and red lines in figure 5.2(a)). Both of them differ from each other in the initial energies and level spacings. The first and second groups, respectively, correspond to the in-phase and out-of-phase subenvelope

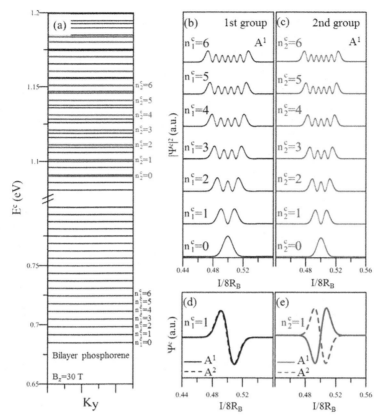

Figure 5.2. For bilayer phosphorene at $B_z = 30$ T, the LL (a) energies, and the probability distributions of the (b) first and (c) second groups on the A^1 sublattice. Also shown are the amplitudes of (d) $n_1^c = 1$ and (e) $n_2^c = 1$ LLs on the A^1 and A^2 sublattices (solid and dashed curves).

functions on A^1 and A^2 (B^1 and B^2) sublattices of two layers, as clearly indicated in figures 5.2(d) and (e).

The highly asymmetric energy dispersion leads to the special dependence of LL energies on $(n^{c,v}, B_z)$, as shown in figures 5.3 and 5.4. In monolayer and bilayer phosphorenes, the low-lying LL energies cannot be described by a simple relation with $n^{c,v} B_z$ (e.g. the linear dependence shown by the dashed lines), especially for a higher energy and field strength. This is different from the square-root dependence in monolayer graphene (figure 3.1(c)) [2], and the linear dependence in AB-stacked graphene (figure 3.7(a); [3, 4]) and MoS_2 (figure 6.2). In general, the LL energies grow with an increment of B_z monotonously. Only the intergroup LL crossings are revealed in the bilayer system (figure 5.4(c)). However, the intragroup and the intergroup anti-crossings are absent, since all the well-behaved LLs are quantized from the monotonous band structure in the energy-wave-vector space (figure 5.1).

The electric field would induce drastic changes in the energy band structures and diverse the magnetic quantizations for bilayer black phosphorus. The first pair of energy bands approaches E_F (figures 5.5(a)–(c)), whereas the opposite is true for the

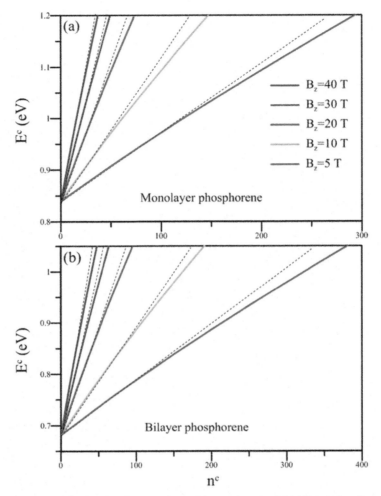

Figure 5.3. The n^c-dependent LL energies in (a) monolayer and (b) bilayer phosphorenes.

second pair. At the critical gate voltage $V_c \approx 2.2$ eV between two layers, the energy gap reduces to zero near the Γ point, as shown in figure 5.5(d). With a further increase in the E-field strength, the low-energy dispersion changes dramatically, as illustrated in figure 5.5(e). Along ΓY and ΓX (unit vectors $\hat{k}_x$ and $\hat{k}_y$), linearly intersecting bands and oscillatory bands exist, respectively. Two split Dirac-cone structures are located at the right- and left-hand sides of the Γ point (along $+\hat{k}_y$ and $-\hat{k}_y$ in figure 5.5(f)). Furthermore, the extremum points are just at the Γ point, accompanied with two saddle points on the opposite k_2s. All the critical points and the constant-energy loops in the energy-wave-vector space will determine the main features of the LL spectra.

The main features of the LLs are dramatically changed for $V_z \geqslant V_c$. The LL spectrum could be divided into three regimes according to the energy ranges of the distinct energy dispersion, e.g. $V_z = 2.4$ eV in figures 5.6(a) and (b). (I), (II), and (III),

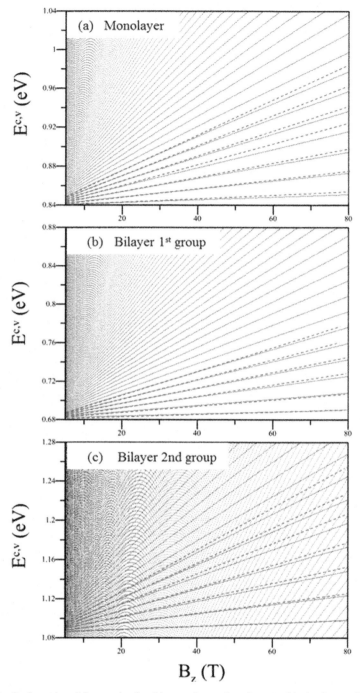

Figure 5.4. The B_z-dependent LL energies for (a) monolayer phosphorene; (b) the first and (c) the second groups in bilayer phosphorene.

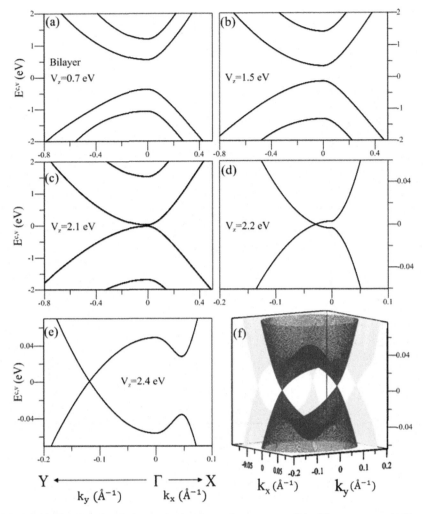

Figure 5.5. (a)–(e) The V_z-induced changes in the electronic structures of the bilayer system. At $V_z = 2.4$ eV, the diverse band dispersions of energy-wave vector space is shown in (f).

respectively, correspond to the Dirac cone (red), the inner and outer parabolic bands (between the saddle and extremum points; green and blue curves), and the parabolic curve (below or above the extremum point, i.e. the blue curve). In (I), the existence of zero-energy LLs at the Fermi level means that the magnetic quantization is initiated from the V_z-induced Dirac point. The two Dirac-cone features make the eigenstate degeneracy of the low-lying LLs twice that of the others, as shown in figure 5.5(f). The LL energy spacings rapidly decrease with increased quantum number. Specifically, the LLs in (II) exhibit an abnormal sequence, a result from the strong competition between magnetic quantization in the two distinct constant-energy loops. The LL spectrum becomes well-behaved in (III), with an almost uniform energy spacing. This directly reflects the normal monotonic quantization of a parabolic band.

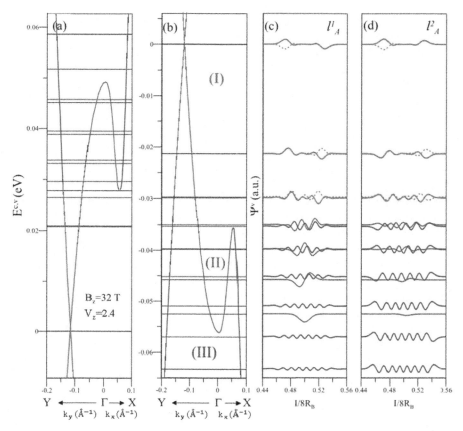

Figure 5.6. At $B_z = 32$ T and $V_z = 2.4$ eV, the (a) conduction and (b) valence LLs, and the subenvelope functions of the latter on the (c) first and (d) second layers. The x-axis in (c) and (d) presents the locations of phosphorus atoms in the enlarged unit cell. This unit cell contains a total of $8R_B$ atoms. Moreover, the different colors used for the curves are associated with the two distinct constant-energy loops and the three energy regions in (b).

The feature-rich LL spectrum can be fully understood from the Landau wavefunctions (figures 5.6(c) and (d)). The subenvelope functions are distributed on to the different sublattices and layers. They are localized near the 1/2 and 2/2 positions of the enlarged unit cell in the crystal lattice space, being associated with the magnetic quantization at the Γ point. These states contribute to the initial LL energy spectra. Similar localization behaviors occur around the two centers, so the 1/2 position is sufficient for a model study. In general, the two sublattices possess identical oscillation modes for the same layer. The number of zero points in the dominant amplitude distribution is the quantum number. There are the coexistent main and side modes for the low-lying LL states at $V_z \geqslant V_c$, a feature due to the combined effects of the multi-hopping integrals, the Coulomb potential energies, and the magnetic field. The subenvelope functions in (I) are simultaneously localized on the left- and right-hand sides of the 1/2 position as shown in figures 5.6(c) and (d), in which the two degenerate states possess the same weight on the two sublattices and layers (light and heavy red curves). They correspond to the magnetic quantization of two neighboring Dirac cones. Such LL states are regarded as the first subgroup of valence LLs. Their zero points in the spatial oscillations increase when the energies are a little away from the Dirac points. Then, the two separate localization centers merge together, which reduces the LL degeneracy to fourfold at $E^v < -0.034$ eV. The zero-point numbers of the l_A^1 and l_A^2 components quickly increase (blue curves) when entering into (II). The well-defined oscillation modes of many zero points clearly demonstrate that such LLs are associated with the quantized states of the larger constant-energy loop. However, the smaller constant-energy loop can induce a second subgroup of LLs (green curves). The initial LLs at -0.053 eV, with the dominant zeroth mode in the l_A^1 component, correspond to magnetic quantization near the Γ point. The competitive quantization between two distinct loops leads to the unfamiliar LL sequence and spatial oscillations in (II). That is, crossings and anti-crossings of two subgroups exist as shown in figures 5.7 and 5.8. The subenvelope functions in (I) and (II) do not present a single-mode oscillation, and their components on the two layers differ from each other. But for (III), they are all well-behaved oscillation modes arising from the monotonic parabolic dispersion, i.e. they are identical to those of a quantized simple harmonic oscillator [5]. Specifically, the LL wavefunctions exhibit the same oscillation modes on two separate layers. In short, the two subgroups of LLs are the magnetically quantized states near the Dirac and Γ points. Their strong competition creates unusual quantization behavior.

The LL energy spectra, shown in figures 5.7 and 5.8, are greatly diversified by applying external fields. The valence and conduction LLs exhibit similar B_z- and V_z-dependent spectra. The former in (I) have a well-behaved B_z-dependence when the magnetic field is smaller than 20 T which can be seen from figure 5.7. Their energies could be fitted by a square-root relation $\sqrt{n_D^v B_z}$. This dependence is similar to that of monolayer graphene with a linear Dirac cone [2]. As for the valence LLs in (II), their spectrum presents two coexistent subgroups for the non-monotonic relations. That is, the n_D^v and n_Γ^v LLs have opposite B_z-dependence. The anti-crossings between the two subgroups of LLs occur when n_D^v and n_Γ^v have quantum number differences $\Delta n = i$. The features clearly illustrate that such LLs are composed of multi-oscillation modes, including a main mode and certain side modes [6]. With the deeper state energies, the LL spectrum is transferred to the monotonic B_z-dependence, directly reflecting the parabolic energy dispersion. All the LLs in (III) belong to single-mode oscillations with many zero points. The complicated LL spectra are also achieved by tuning the electric field. At $V_z > V_c$, for which we refer to figure 5.8, frequent crossings and anti-crossings related to the two subgroups of LLs exist. The diverse LL spectra will be obviously revealed in DOSs by special structures, so that they could be directly identified from the STS measurements.

The main features of the energy bands and LL spectra are directly reflected in the DOS. In zero fields or in a weak electric field, the band-edge states of parabolic bands create gap-dependent shoulder-like structures, e.g. $V_z = 2.1$ eV in figure 5.9(a) shown as the black curve. When the gap transition occurs at $V_z = V_c$ (figure 5.5(e)),

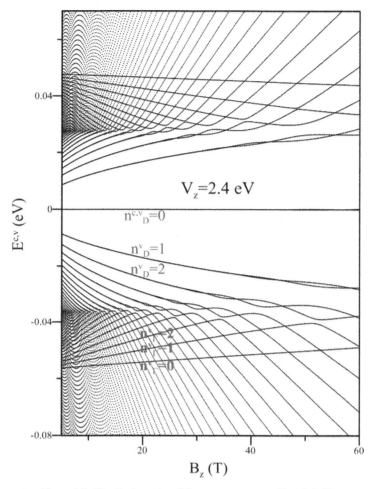

Figure 5.7. The B_z-dependent LL energy spectra at $V_z = 2.4$ eV.

the initial structures are replaced by a valley-like structure due to the deformed Dirac cone (figure 5.9(b)). The logarithmically divergent peaks (the symmetric ones), corresponding to the saddle points in energy bands, come to exist on both sides of this valley. The special structures are more obvious when $V_z > V_c$ (figures 5.9(c) and (d)). Furthermore, shoulder structures exist arising from the extremum Γ points. The prominent peaks and shoulders gradually move away from the Fermi energy with a further increase in V_z. The V_z-induced drastic changes in the DOS could be verified by STS measurements.

A uniform magnetic field induces several delta-function-like peaks. The height and spacing of the peaks reflect the eigenstate degeneracy and the energy dispersion of $B_z = 0$. At $V_z < V_c$, the low-frequency DOS peaks have uniform height with four-fold degeneracy and almost the same spacing as shown as the magenta curve in figure 5.9(a), mainly due to the quantization of the parabolic energy dispersion (see

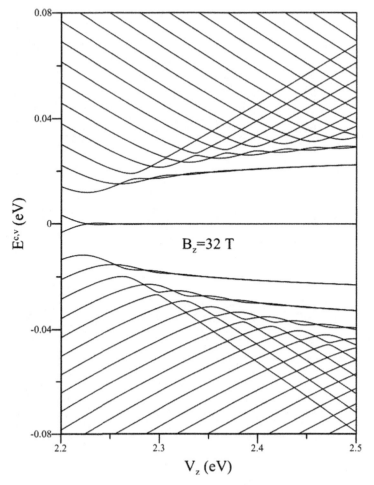

Figure 5.8. The V_z-dependent LL energy spectra at B_z = 32 T.

figure 5.5(c)). There is a pair of peaks centered around the Fermi level at $V_z = V_c$ in figure 5.9(b). With an increase in V_z, a very prominent peak, with eight-fold degeneracy, appears at $E_F=0$ as shown in figure 5.9(c). Similar peaks, which come from the quantized Dirac cone, could survive at stronger electric fields as can be seen in figure 5.9(d). The double-peak character at higher energies is due to two anti-crossing LLs. Apparently, all the low-lying peaks exhibit highly non-uniform spacings. The STS measurements on the main features of low-energy LL peaks could provide useful information about the diverse magnetic quantizations.

Many critical features of the (magneto) electronic properties of phosphorene can be further validated by various measurements. The peculiar wave-vector-dependent energy dispersion and electrically-tunable band gap may be measured by ARPES [7, 8]. Transport measurements are shown to be capable of estimating the energy gap or even gate-voltage-dependent gap [9]. The optical measurements may be used to

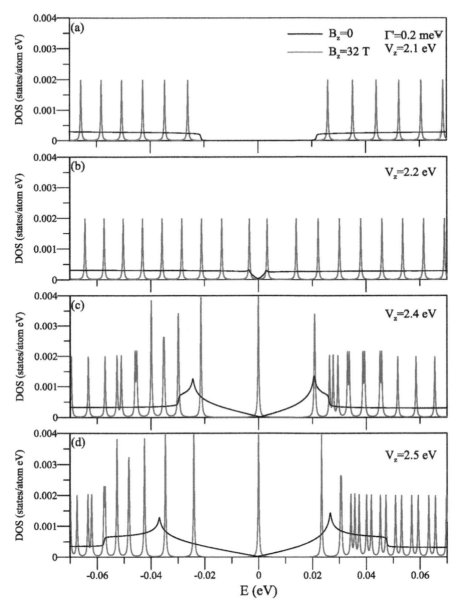

Figure 5.9. DOSs of bilayer phosphorus at $B_z = 0$ and 32 T (black and magenta curves) (a)–(d) under various V_z's. $\Gamma' = 0.2$ meV is the broadening factor in calculating DOS.

characterize the band gap or inter-band-edge state excitations [10]. Magneto-optical measurements could be also performed to examine the inter-LL excitations. Recently, quantum Hall plateaus in phosphorene are clearly observed, as reported in [11]. It can be predicted that the strong E_z effects on the LL spectrum may alter the sequence of Hall plateaus significantly.

References

[1] Rudenko A N and Katsnelson M I 2014 Quasiparticle band structure and tight-binding model for single- and bilayer black phosphorus *Phys. Rev.* B **89** 201408

[2] Ho J H, Lai Y H, Chiu Y H and Lin M F 2008 Landau levels in graphene *Physica* E **40** 1722–5

[3] Lin C Y, Wu J Y, Chiu Y H and Lin M F 2014 Stacking-dependent magneto-electronic properties in multilayer graphenes *Phys. Rev.* B **90** 205434

[4] Lin C Y, Wu J Y, Ou Y J, Chiu Y H and Lin M F 2015 Magneto-electronic properties of multilayer graphenes *Phys. Chem. Chem. Phys.* **17** 26008–35

[5] Lai Y H, Ho J H, Chang C P and Lin M F 2008 Magnetoelectronic properties of bilayer Bernal graphene *Phys. Rev.* B **77** 085426

[6] Wu J Y, Chen S C, Gumbs G and Lin M F 2017 Field-induced diverse quantizations in monolayer and bilayer black phosphorus *Phys. Rev.* B **95** 115411

[7] Ehlen N, Senkovskiy B V, Fedorov A V, Perucchi A, di Pietro P, Sanna A, Profeta G, Petaccia L and Grüneis A 2016 Evolution of electronic structure of few-layer phosphorene from angle-resolved photoemission spectroscopy of black phosphorous *Phys. Rev.* B **94** 245410

[8] Kim J *et al* 2015 Observation of tunable band gap and anisotropic Dirac semimetal state in black phosphorus *Science* **349** 723

[9] Liu H, Neal A T, Zhu Z, Luo Z, Xu X, Tománek D and Ye P D 2014 Phosphorene: an unexplored 2D semiconductor with a high hole mobility *ACS Nano* **8** 4033

[10] Qiao J, Kong X, Hu Z X, Yang F and Ji W 2014 High-mobility transport anisotropy and linear dichroism in few-layer black phosphorus *Nat. Commun.* **5** 4475

[11] Li L *et al* 2016 Quantum Hall effect in black phosphorus two-dimensional electron system *Nat. Nanotechnol.* **11** 593

Chapter 6

MoS$_2$

MoS$_2$ sharply contrasts with group-IV systems in terms of lattice symmetries, atomic hopping integrals and SOCs, being further reflected in the essential electronic properties. It can present a unique band structure with a middle energy gap, orbital-dependent spin splitting, and multi-valleys (centered at the K,K' and Γ points). Such features are due to the Mo-($4d_{x^2-y^2}$,$4d_{xy}$,$4d_{z^2}$)-dominated site energies and SOCs. The novel magnetic quantization lies in the coexistence of the K/K'-valley- and spin-split LL subgroups. Four LL subgroups exist during the variation of B_z, being absent in the Γ-valley-initiated quantum states. The energy splittings are gradually enlarged by the increasing B_z. The combined effects arising from the intrinsic properties and the external field are proposed to account for the diversified phenomena. Such LLs are doubly degenerate, in which the well-defined oscillation modes in the B_z-dependent energy spectra exhibit non-crossing or crossing behavior.

Layered MoS$_2$ is very different from the inorganic group-IV systems in lattice structures, orbital hybridization, and SOC, leading to unusual essential properties. For a MoS$_2$ monolayer, the three-orbital tight-binding model is sufficient to describe the electronic properties. In the bases of $\{|4d_{z^2}\rangle, |4d_{xy}\rangle, |4d_{x^2-y^2}\rangle\} \otimes \{\uparrow, \downarrow\}$, the Hamiltonian is

$$
\begin{aligned}
H = &\sum_{\langle I\rangle,o,m} E_o C_{Iom}^+ C_{Iom} \\
&+ \sum_{\langle I,J\rangle,o,o',m} \gamma_{oo'}^{\Delta \mathbf{R}_J} C_{Iom}^+ C_{Jo'm} + \sum_{\langle I\rangle,o,o',m} \frac{\lambda_{\mathrm{SOC}}}{2} C_{Iom}^+ C_{Io'm} \left(L_{z,oo'}\sigma_{mm}^z \right),
\end{aligned}
\tag{6.1}
$$

where the first, second and third terms are, respectively, the site energy, the nearest-neighbor hopping integral and the on-site SOC. The site energies are distinct for the $4d_{z^2}$ and ($4d_{xy}$,$4d_{x^2-y^2}$) orbitals, and this difference will induce the valley-dependent LLs. The SOC is only contributed by the z-component angular momentum (L_z) and spin moment (σ^z). Such interaction occurs between $|4d_{x^2-y^2}\rangle$ and $|4d_{xy}\rangle$ with the same spin configuration, while it is independent of $|4d_{z^2}\rangle$. All the critical interactions in equations (5.1) and (3.3) sharply contrast with each other.

The multi-orbital bondings and the SOC cause monolayer MoS_2 to exhibit an unusual electronic structure. A direct energy gap of 1.59 eV at the K or K' point, as

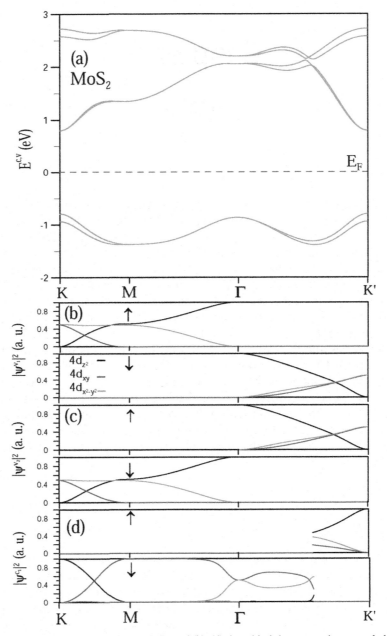

Figure 6.1. (a) Energy bands of monolayer MoS_2, and (b)–(d) the orbital-decomposed state probabilities along the high-symmetry points.

shown in figure 6.1(a), is dominated by the site energies of distinct orbitals. The significant orbital hybridizations lead to the strong wave-vector dependence. The electronic states of parabolic bands near the Fermi level are centered at the K, K' and Γ points. Furthermore, the SOC can create the spin-split energy bands, e.g. the largest splitting energy is $2\lambda_{SOC}$ at the K and K' points. Whether split spin-up and spin-down energy bands exist is dependent on the components of $4d_{x^2-y^2}$ and $4d_{xy}$ orbitals. The contributions of these two orbitals, as indicated by Ψ^{v_1} and Ψ^{v_2} in figures 6.1(b) and (c), are comparable in the split valence bands near the K and K' points. However, when the electronic states mainly come from one of them, or the $4d_{z^2}$ orbital, the negligible SOC makes the spin splitting almost vanish, e.g. the lower-lying conduction bands in figure 6.1(d). It should also be noted that the group-IV systems do not have spin-split energy bands as a result of the mirror symmetry in A and B sublattices (figures 4.1(a) and 4.5(a)).

MoS$_2$ systems present a unique magnetic quantization, since the valley- and spin-dependent LL subgroups can survive simultaneously. All the LLs have two degenerate localization centers, the 1/2 and 2/2 localization centers in an enlarged unit cell, e.g. the 1/2 localized LL wavefunctions shown in figures 6.1(b) and 6.3(b). That is, LLs remain the same under the simultaneous interchanges in the localization center, valley and spin. The dominating modes have well-behaved spatial probability distributions. Each mode is fully determined by the spin-up or spin-down configuration, but not a superposition of two opposite spins, as revealed in group-IV systems (figures 4.2(b), 4.2(d), 4.6(b) and 4.6(d)). Each LL group of monolayer MoS$_2$ only corresponds to the occupied LLs or the unoccupied LLs, while that in group-IV systems includes those for valence and conduction (e.g. figure 3.5(c), figure 3.7 and figure 4.3(c)). It is further split into LL subgroups under the destruction of the spin and/or valley degeneracy.

As to the valence LLs (figure 6.2(a)), they are magnetically quantized from the electronic states centered at the Γ, K and K' points (figure 6.1(a)). The Γ-dependent LL wavefunctions are independent of the spin configuration ($n_{\Gamma\downarrow}^v = 0$, $n_{\Gamma\uparrow}^v = 0$, $n_{\Gamma\downarrow}^v = 1$, and $n_{\Gamma\uparrow}^v = 1$ in figure 6.2(b)), mainly owing to the $4d_{z^2}$-orbital dominance and the almost zero SOC. The LL energies linearly grow with B$_z$ (blue curves in figure 6.2(a)), directly reflecting the parabolic dispersion near the Γ point. A similar B$_z$-dependence is revealed in the energy spectra of other LL subgroups. However, the spin-up and spin-down (spin-down and spin-up) LL subgroups that come from the K(K') valley are initiated at −0.792 eV and −0.938 eV, respectively, (black curves in figure 6.2(a); yellow and red between figures 6.2(a) and (b)). The spin-split LL subgroups are closely related to the $4d_{x^2-y^2}$- and $4d_{xy}$-dominated SOCs, as indicated in figures 6.1(b) and (c). Specifically, the degeneracy of the two valleys is clearly destroyed by an increase of B$_z$. That is to say, K- and K'-dependent LL subgroups also exist (solid and dashed lines in yellow or red). The energy spacing is observable for a sufficiently high B$_z$, e.g. ∼15 meV between the $n_{K\uparrow}^v=0$ and $n_{K'\downarrow}^v = 0$ LLs at B$_z$=40 T. A detailed analysis shows that the site-energy differences in the B$_z$-enlarged unit cell is responsible for these LL subgroups [1]. The coexistence of the spin- and valley-dependent LL subgroups is absent in group-IV systems even under a

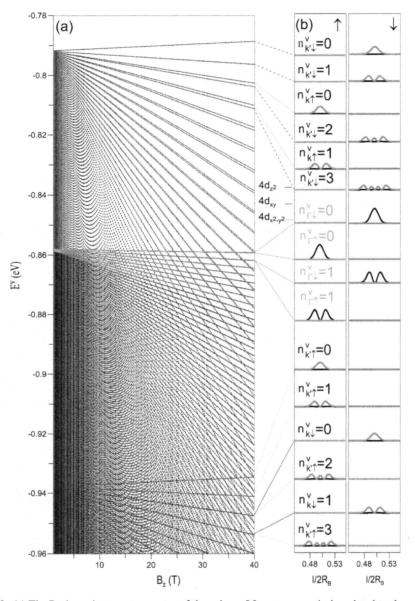

Figure 6.2. (a) The B_z-dependent energy spectra of the valence LLs are, respectively, related to the quantized states near the (K,K') and Γ points (black and blue curves), in which the valley-dependent (spin-dependent) subgroups are represented by the solid and dashed curves (yellow and red). The spatial probability distributions are shown in (b) at $B_z = 40$ T.

composite magnetic and electric field. Such subgroups do not cross or anticross with one another, while they might have crossings with Γ-dependent ones (black and blue curves in figure 6.2(a)). However, the lower-lying conduction LLs, as shown in figures 6.3(a) and (b), possess significantly K- and K'-dependent subgroups (solid

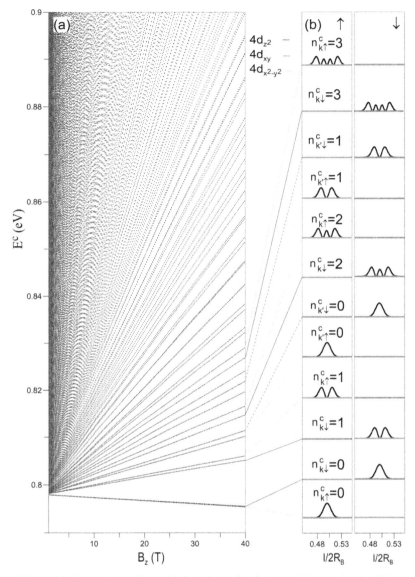

Figure 6.3. Same plot as figure 6.2, but shown for the lower-lying conduction LLs.

and dashed lines), and weak splittings in the spin-dependent subgroups (yellow and red). Furthermore, they present a linear B_z-dependent energy spectrum without crossings and anti-crossings, corresponding to the parabolic conduction bands near the K and K' points (figure 6.1(a)).

The zero field band structures of the layered transition metal dichalcogenides have been validated by the measurements of ARPES [2–5]. The thickness-dependent electronic structure verifies the transition from indirect to direct band gap in monolayers. This transition is also observed by the STS measurement [6–8] and the optical spectroscopy [9, 10]. The Shubnikov–de Haas oscillations have been observed in an hBN-encapsulated MoS_2 device [11]. Magneto-optical measurements could be used to examine the inter-LL excitations between the spin- and valley-dependent LLs.

The unusual magneto-electronic properties in the layered systems could be further verified using STS, including the normal and abnormal B_z-dependences in ABC-stacked graphenes, three kinds of LLs in sliding stacking systems, SOC-induced spin-dominated LLs in germanene and silicene, two groups of low-lying LLs in tinene, spin- and valley-dependent LLs in MoS_2, special $n^{c,v}$- and B_z-dependence of LL energies in few-layer phosphorenes, and the B_z- and V_z-dependent energy spectra with the LL splittings, crossings and anti-crossings. STS examinations can provide critical information on the lattice symmetry, layer number, stacking configuration, SOC, single- or multi-orbital hybridization and the external fields.

References

[1] Ho Y H, Wang Y H and Chen H Y 2014 Magnetoelectronic and optical properties of a MoS_2 monolayer *Phys. Rev.* B **89** 55316

[2] Jin W *et al* 2013 Direct measurement of the thickness-dependent electronic band structure of MoS_2 using angle-resolved photoemission spectroscopy *Phys. Rev. Lett.* **111** 106801

[3] Zhang Y *et al* 2014 Direct observation of the transition from indirect to direct bandgap in atomically thin epitaxial $MoSe_2$ *Nat. Nanotechnol.* **9** 111–5

[4] Cheng M K *et al* 2017 Large-area epitaxial growth of $MoSe_2$ via an incandescent molybdenum source *Nanotechology* **28** 455601

[5] Tang S *et al* 2017 Quantum spin Hall state in monolayer 1T'-WTe_2 *Nat. Phys* **13** 683

[6] Huang Y L *et al* 2015 Bandgap tunability at single-layer molybdenum disulphide grain boundaries *Nat. Commun.* **6** 6298

[7] Park J H *et al* 2016 Scanning tunneling microscopy and spectroscopy of air exposure effects on molecular beam epitaxy grown WSe2 monolayers and bilayers *ACS Nano* **10** 4258

[8] Zhang Y *et al* 2016 Electronic structure, surface doping, and optical response in epitaxial WSe_2 thin films *Nano Lett.* **16** 2485

[9] Castellanos-Gomez1 A, Quereda J, van der Meulen H P, Agraït N and Rubio-Bollinger G 2016 Spatially resolved optical absorption spectroscopy of single- and few-layer MoS_2 by hyperspectral imaging *Nanotechnology* **27** 115705

[10] Ottaviano L *et al* 2017 Mechanical exfoliation and layer number identication of MoS_2 revisited *2D Mater* **4** 045013

[11] Cui X *et al* 2015 Multi-terminal transport measurements of MoS_2 using a van der Waals heterostructure device platform *Nat. Nanotechnol.* **10** 534–40

Chapter 7

Non-uniform magnetic fields in graphene

In addition to a uniform magnetic field, generalized tight-binding is also suitable in studying non-uniform quantization properties due to a periodic field, or a composite one. The former in monolayer graphene can create lowly degenerate QLLs with strong energy dispersions and high anisotropy. The 1D characteristics might lead to a lot of square-root-form asymmetric peaks in DOS. The magnetic wave functions will present distorted spatial distribution and even a transformation of oscillation modes, depending on the strength and period of a spatially modulated field. This field has very strong effects on the main features of well-behaved LLs, covering the enhancement in dimensionality, reduction of state degeneracy, variation of energy dispersions, creation of band-edge states, and the changes in the center, width, phase and symmetric/anti-symmetric distribution of the localized quantum modes.

7.1 A spatially modulated magnetic field

The theoretical calculations for uniform, non-uniform and composite magnetic fields are discussed in section 2.1. A monolayer graphene, as shown in figure 7.1, is assumed to exist in a spatially sinelike-modulated magnetic field $\mathbf{B}_M = B_M \sin(2\pi x/l_M)\,\hat{z}$ along the armchair direction ($\hat{x}$). A primitive unit cell, corresponding to the periodic length l_M ($3bR_M$), includes $2R_M$ A atoms and $2R_M$ B atoms. The magneto-electronic structure built from the $2p_z$ orbitals is characterized by $4R_M$ tight-binding functions. The Peierls phases in the neatest-neighbor hopping integrals are associated with the vector potential $\mathbf{A} = -B_M l_M \cos(2\pi x/l_M)/2\pi\,\hat{y}$. The analytic formulas are similar to those in equations (2.3)–(2.5). It is relatively easy to calculate electronic structures, wave functions and DOS in the band-like Hamiltonian matrix (details in [1–4]).

We focus on the low-energy electronic properties modulated by non-uniform magnetic fields. Monolayer graphene in the presence of $B_M = 8$ T and $R_M = 1000$ is chosen for a model study, in which the dependence of electronic properties on k_x is very weak and negligible. At zero field, a pair of Dirac cones near the K point

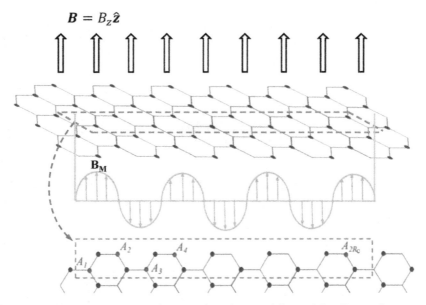

Figure 7.1. Geometric structure of monolayer graphene in a spatially modulated/composite magnetic field along the armchair direction. The uniform and modulated magnetic fields are, respectively, denoted by **B** and **B**$_M$. An enlarged unit cell corresponds to a composite field.

(figure 3.1(a)) is folded into many subbands centered at the original band-edge state $k_y^o = 2/3$ (in unit of $\pi/\sqrt{3}\,b$) under an enlarged unit cell, as shown in figure 7.2 by the black solid circles. Most of energy subbands present parabolic dispersions with double degeneracy except for the two non-degenerate linear bands intersecting at E_F. There is only one band-edge state in each energy subband; furthermore, all the band-edge states are located at k_y^o. A modulated magnetic field does not alter the symmetry of the band structure about E_F, while it induces dramatic changes in energy dispersions and band-edge states (open blue circles). The k_y-range of the low-lying electronic states becomes wide. The linear subbands are changed into partial flat subbands at E_F, as observed for carbon nanotubes in magnetic fields perpendicular to the symmetry axis [5]. The number of doubly degenerate parabolic subbands is greatly reduced. Such subbands have weak energy dispersions or low curvatures near k_y^o, in which their energies approach those of the LLs in a uniform magnetic field (the dashed red lines). These results indicate that a magnetic field could make the neighboring electronic states flock together. Moreover, the modulation effects on parabolic subbands lead to four extra band-edge states at k_{be}'s, strong energy dispersions close to them, and the destruction of the double degeneracy. The two extra band-edge states at the left- and right-hand sites of k_y^o might have distinct energies; that is, one side of the parabolic subbands might be asymmetric to the other about the original band-edge states. Each parabolic subband exhibits composite behavior in state degeneracy, the single and double degeneracies near k_{be} and k_y^o, respectively.

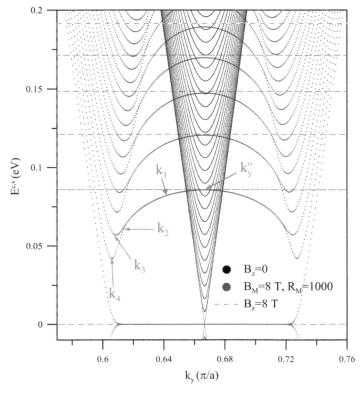

Figure 7.2. The low-lying band structure of monolayer graphene in a periodic magnetic field at $B_M = 3$ T and $R_M = 1000$. Also shown for comparison are those at zero field (solid circles) and a uniform magnetic field $B_z = 8$ T (the red dashed lines). The different k_y's are indicated by arrows.

Each energy subband, which is almost dispersionless near k_y^o, could be regarded as a quasi-Landau level, as clearly indicated from the main characteristics of the wave functions in figure 7.3. The odd- and even-index carbon atoms present the same or opposite phases in wave functions, in agreement with those under a uniform magnetic field. As for the lowest unoccupied QLL of $n_Q^c = 0$, the wave function at k_y^o possesses a well-behaved distribution mode without zero point, being similar to $n^c = 0$ LL (figure 3.1). It is, respectively, localized at one fourth ($x_1 = 3/4$) and three fourths ($x_2 = 1/4$) of a unit cell in the A and B sublattices (the solid black circles in figures 7.3(a) and (b)). x_1 and x_2 correspond to the maximum magnetic-field strength, meaning the strongest quantization effect. However, the maximum amplitude is different for these two sublattices, indicating the destruction of the sublattice equivalence by a modulated magnetic field. A similar spatial distribution is revealed in the highest occupied QLL of $n_Q^v = 0$ (the open blue circles in figures 7.3(a) and (b)); that is, the $n_Q^v = 0$ and $n_Q^c = 0$ QLLs have identical probability distribution under the interchange of two sublattices and localization positions. In addition, the width of probability distribution at half maximum is comparable to the

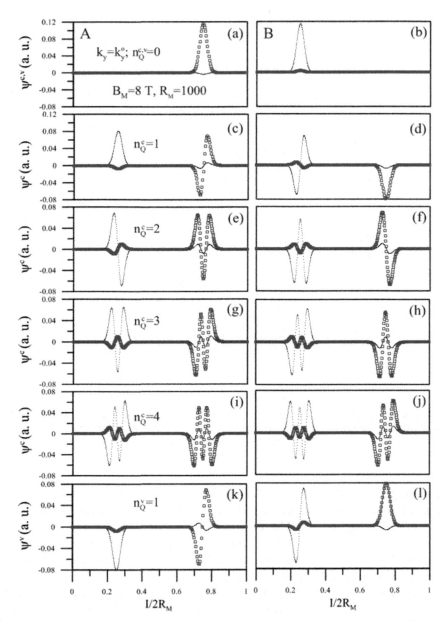

Figure 7.3. The wave functions in the A and B sublattices of a unit cell for the QLLs: (a)–(b) $n_Q^{c,v} = 0$, (c)–(d) $n_Q^c = 1$, (e)–(f) $n_Q^c = 2$, (g)–(h) $n_Q^c = 3$, (i)–(j) $n_Q^c = 4$; (k)–(l) $n_Q^v = 1$. Black solid and blue open circles represent two degenerate states.

magnetic length $\sqrt{\hbar/eB_M}$ (~70 Å at $B_M = 8$ T), being much longer than the original lattice constant in the honeycomb lattice.

In sharp contrast with the $n_Q^c = 0$ QLL, the second unoccupied QLLs are doubly degenerate at k_y^o; furthermore, their wave functions consist of two kinds of localization modes centered at x_1 and x_2, as clearly shown in figures 7.3(c) and (d). Ψ^c, respectively, has no and one zero point in the A sublattice, while the opposite is true for the B sublattice. The effective quantum number is characterized by the larger number of zero point, i.e. $n_Q^c = 1$ is used to define the second unoccupied QLLs. In addition, the two-fold degenerate QLLs possess the same probability distributions, in which their difference only lies in the sign change of the subenvelope functions. The other unoccupied and occupied QLLs exhibit the similar quantum modes at k_y^o (figures 7.3(e)–(l)). For example, the n_Q^c QLLs have $n_Q^c - 1$ and n_Q^c zero points for the A sublattice at x_1 and x_2, respectively. It can be deduced that the essential properties due to the QLLs near k_y^o are almost identical to those from the LLs, such as the special structures in DOS and the magneto-optical selection rule [4].

The QLL wave functions, as shown in figures 7.4(a)–(l), would be strongly modified when the wave vector gradually moves away from the original band-edge

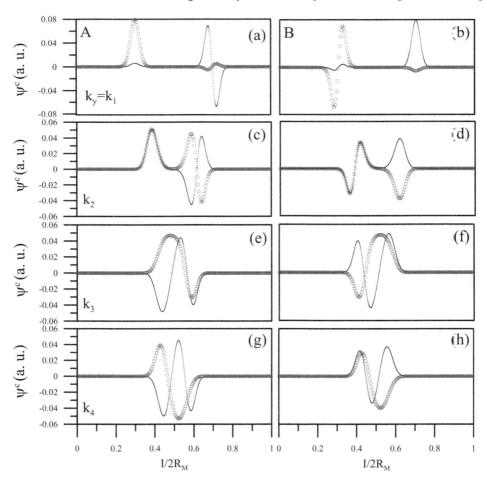

Figure 7.4. Similar plots as in figure 7.3, but shown for the second conduction QLLs at different k_y's: (a)–(b) k_1, (c)–(d) k_2, (e)–(f) k_3, and (g)–(h) k_4. The lower- and higher-subband states are, respectively, denoted by black solid and blue open circles.

state (k_y^o). They present similar features for the left- and right-side wave vectors, so the former are chosen to illustrate the main characteristics. For the second unoccupied QLLs with double degeneracy at $k_y = k_1$ (figure 7.2), their wave functions are centered at $x_1 = 3/10$ and $x_2 = 7/10$ (figures 7.4(a) and (b)). The spatial oscillation modes are almost the same with those at k_y^o (figures 7.3(a) and (b)), while the distance between the two localization centers ($|x_1 - x_2|$) is, respectively, 2/5 and 1/2 for the former and the latter. At $k_y = k_2$ (figure 7.2), the doubly degenerate QLLs start to separate into two non-degenerate ones. The localization centers ($x_1 \sim 7/20$ and $x_2 \sim 13/20$) are sufficiently close so that $\Psi^c(x_1)$ and $\Psi^c(x_2)$ have a slight overlap (solid black/open blue circles in figures 7.4(c) and (d)). Specifically, k_3 and k_4 are, respectively, the band-edge states of the higher and lower subbands (figure 7.2). They have similar spatial distributions, as indicated in figures 7.4(e) and (f) and figures 7.4(g) and (h). For example, the second unoccupied QLLs at k_3 behave as two non-degenerate subbands, in which the wave function of the lower subband exhibits a strong overlap and thus irregular oscillation modes (blue open circles in figures 7.4(e) and (f)). Apparently, the two localization centers are merged together and approach half of a unit cell; that is, their positions correspond to almost zero field strength. In other words, most of the probability density is transformed from the region near the highest magnetic field to that of the lowest one. The drastic changes in the distribution modes and probability densities are also revealed in the k_3 state of the higher subband (black solid circles in figures 7.4(e) and (f)) and the k_4 state in two non-degenerate subbbands (figure 7.4(g) and (h)). According to the above-mentioned results, the main difference between QLLs and LLs lies in the abnormal/well-behaved quantum modes near the extra band-edge states. This will be directly reflected in other essential properties, e.g. magneto-optical selection rules [4].

The band-edge states of QLLs can create a lot of special structures in DOS, as clearly shown in figures 7.5(a) and (b). A delta-functional-like prominent peak and many asymmetric peaks in the square-root form exist, which is different from the linear E-dependence due to the Dirac cone at zero field (the black solid line in figure 7.5(a)). The former comes from a pair of partial flat subbands merged at E_F (figure 7.2), and its height grows with the magnetic field strength. The latter are due to the extreme points of the 1D parabolic dispersions along $\hat{k}_y$, in which their intensities are inversely proportional to the subband curvatures. The asymmetric peaks could be further divided into strong principal peaks and weak subpeaks, being induced by the original and extra band-edge states, respectively. The asymmetric/symmetric strong peaks become symmetric (the red dashed curve), when a modulated magnetic field is replaced by a uniform one. The number, frequencies, and heights of the peak structures are sensitive to the changes in strength, period, and modulation direction. The peak number declines with increasing field strength (figure 7.5(a)), while the peak frequencies present an obvious enhancement. The number of asymmetric subpeaks grow in the increment of the period (figure 7.5(b)). However, the main features of the principal peaks weakly depend on the period. In addition, the DOS is highly anisotropic, as illustrated by the modulation directions along the zigzag and

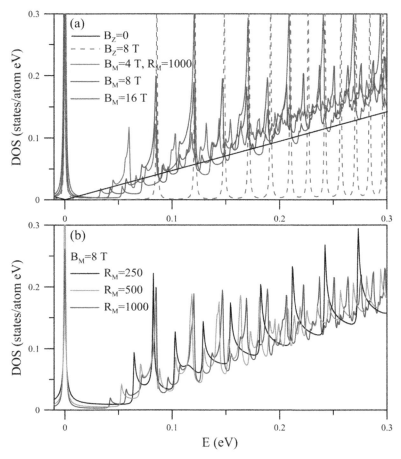

Figure 7.5. DOSs of QLLs in the spatially modulated magnetic fields along the armchair direction under (a) various field strengths for a specific period $R_M = 1000$ and (b) different periods at field strength $B_M = 8$ T. Also shown are those in (a) for zero field and a uniform magnetic field $B_z = 8$ T.

armchair structures (details in [2, 4]). More principal peaks, with lower frequencies, are revealed in the zigzag case.

The modulated magnetic and electric fields can greatly diversify the essential properties [2, 6–11]. However, strong effects on the electronic properties are quite different between them. For the Hamiltonian of monolayer graphene, the former and the latter, respectively, induce Peierls phases in the hopping integrals and distinct site energies. A modulated electric potential is unable to create the magnetic quantization phenomena; therefore, the QLLs with the localized probability distributions, are replaced by a lot of 1D subbands with oscillatory, parabolic or linear energy dispersions. Each electronic state presents a standing wave in a unit cell; furthermore, its distribution width is much wider than that of a localized QLL. In particular, this potential could make a semiconducting graphene become a semi-metallic system, since DOS near the Fermi level changes from zero into a finite value. Apparently, an electric field cannot generate the optical selection rules being observed in a magnetic field.

7.2 A composite magnetic field with uniform and modulated components

Monolayer graphene is present in a composite magnetic field with uniform $B_z\hat{z}$ and modulated $B_M\sin(2\pi x/l_M)\hat{z}$ along the armchair structure, as shown in figure 7.1. These two components, respectively, induce the periods R_B and R_M; therefore, the least common multiple R_C corresponds to an enlarged unit cell (figure 7.1). The Hamiltonian matrix is a $4R_C \times 4R_C$ Hermitian matrix spanned by $4R_C$ tight-binding functions, as for a uniform magnetic field. The independent matrix elements are similar to those in equation (2.5) (details in [3, 12]). The effects of a modulated magnetic field on highly degenerate LLs could be explored in detail by gradually enhancing its strength. For a large R_C, where the range of k_x is much smaller than that of k_y, it is sufficient only to consider the energy dispersions along $\hat{k}_y$ at $k_x = 0$ in the following calculations.

A spatially modulated magnetic field leads to drastic changes of LLs in energy dispersions, state degeneracies and band-edge states, as clearly shown in figures 7.6 and 7.7. We first discuss the modulation effects due to a composite magnetic field under $B_z = 32$ T, $B_M = 4$ T and $R_c = R_B = R_M$ (figure 7.6(a)). Monolayer graphene has a lot of four-fold degenerate LLs with well-separated energy spacings (the dashed red lines; details in section 3.1). All the dispersionless LLs become 1D parabolic subbands, except that the Landau level of $n^{c,v} = 0$ at E_F is independent of the modulated magnetic fields. Such energy subbands could be characterized as QLLs because of the spatial quantization modes (discussed later in figures 7.8 and 7.9). The stronger energy dispersions indicate that the wave functions of the LLs will be greatly modified by a non-uniform magnetic quantization; furthermore, one dimensionality is recovered by it. Their band widths grow with increasing state energies. The fourfold degeneracy is completely destroyed, in which four non-degenerate parabolic subbands present the crossing and overlapping behaviors, and each subband possesses one band-edge state at $k_y \neq 0$, e.g. k_{be} of the lowest conduction subband (a red circle in figure 7.6(a)). In addition, the non-degenerate conduction subbands are denoted as c_1, c_2, c_3 and c_4.

The main features of 1D energy subbands strongly depend on the strength and period of a modulated magnetic field. The modulation strength, as clearly shown in figures 7.6(a)–(c), would significantly alter the subband curvature (inversely proportional to the effective mass) and shift the band-edge state energies. The stronger B_M, the larger the subband curvature/subband overlap. The energies of local maxima (minima) grow (decline) with an increment of B_M. The number of band-edge states remains the same during the variation of B_M. In particular, the parabolic subbands are further changed into valley-like ones with unusual curvatures near the same band-edge state ($k_{be} = 1/3$ for $B_M = B_Z$ in figure 7.6(d)), when the modulated and uniform magnetic fields are comparable. This indicates a thorough destruction of the LL characteristics. As for the modulation period, it presents diverse effects

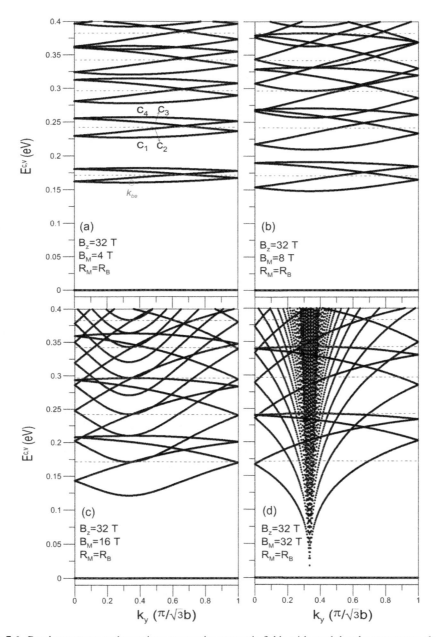

Figure 7.6. Band structures under various composite magnetic fields with modulated components along the armchair direction ($\hat{x}$) for B_z=32 T, $R_M = R_B$ and (a) $B_M = 4$ T, (b) $B_M = 8$ T (c) $B_M = 16$ T and (d) $B_M = 32$ T. Also shown by the red dashed lines are those of the LL energies at $B_z = 32$ T. k_{be} (a red circle) corresponds to the band-edge state of the lowest conduction subband of $n_Q^c = 1$. ($c_1 c_2 c_3 c_4$) represent the moculation-induced non-degenerate four conduction subbands.

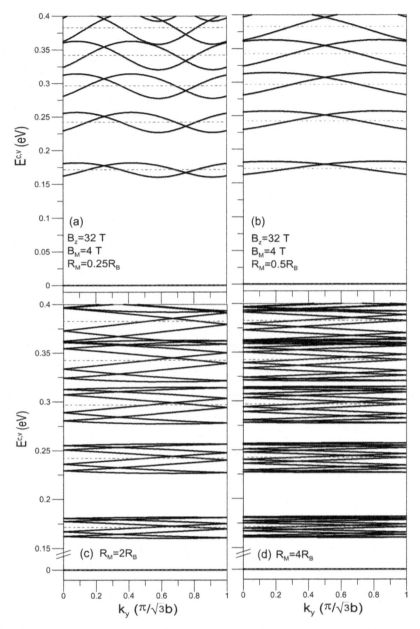

Figure 7.7. Similar plots as in figure 7.6, but shown for $B_z = 32$ T, $B_M = 4$ T and (a) $R_M = R_B/4$, (b) $R_M = R_B/2$, (c) $R_M = 2R_B$ and (d) $R_M = 4R_B$.

according to the ratio of $R = R_M/R_B$. The number of energy subbands in the first Brillouin zone is fixed for $R < 1$ (figures 7.7(a) and (b)), while it is proportional to R for $R > 1$ (figure 7.7(c) and (d)). The doubly degenerate states come to exist in the former case when $1/R$ is an even integer. On the contrary, the band curvatures

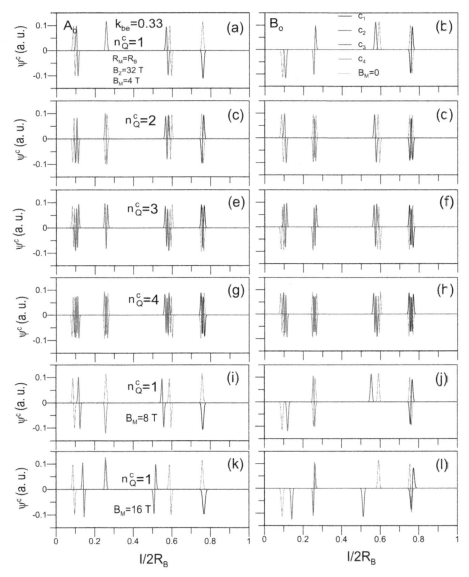

Figure 7.8. The wave functions of the various conduction QLLs at $k_{be} = 0.33$ under a composite magnetic field with $R_M = R_B$ and $B_z = 32$ T: $B_M = 4$ T for (a)–(b) $n_Q^c = 1$, (c)–(d) $n_Q^c = 2$, (e)–(f) $n_Q^c = 3$; (g)–(h) $n_Q^c = 4$, and (i)–(j) $B_M = 8$ T; (k)–(l) $B_M = 16$ T for $n_Q^c = 1$. Also shown by the dashed curves are those of the four-fold degenerate LLs.

increase for $R < 1$, whereas they almost remain the same for $R > 1$. It should be noted that the band-edge state energies (the subband widths) hardly depend on the modulation period.

The LL wave functions are strongly affected by modulated magnetic fields. The magnetic localization centers are very sensitive to a perturbed field. Their positions

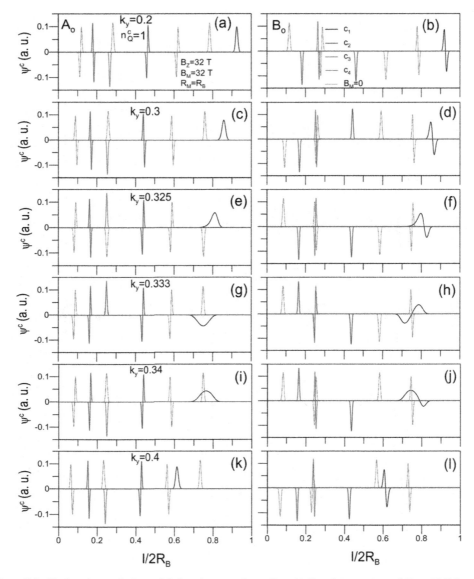

Figure 7.9. Similar plots as in figure 7.8, but shown at $B_M = B_z = 32$ T and $R_M = R_B$ and $B_z = 32$ T for the $n_Q^c = 1$ electronic states: (a)–(b) $k_y = 0.2$, (c)–(d) $k_y = 0.3$, (e)–(f) $k_y = 0.325$, (g)–(h) $k_y = k_{be} = 0.333$, (i)–(j) $k_y = 0.34$, and (k)–(l) $k_y = 0.4$.

exhibit obvious changes even under a weak modulation, as clearly shown in figures 7.8(a)–(h) at $B_z = 32$ T, $B_M = 4$ T and $R_M = R_B$. For the n^c conduction LLs with four-fold degeneracy, there are four distinct localization centers in a uniform magnetic field (the dashed curves; section 3.1). The first, second, third and fourth centers, respectively, present the n^c, $n^c - 1$, n^c and $n^c - 1$ quantum modes in the A sublattice, in which the two distinct modes are interchanged in the B sublattice.

Furthermore, the neighboring distances are 1/6, 2/6 and 1/6, respectively. Such LLs become non-degenerate QLLs in a modulated magnetic field. The wave functions of the latter are, respectively indicated by the black, blue, red and green curves (the fourth, third, first and second localization centers) according the ordering of the split energies (figure 7.6(a)). The first and third localization centers in an enlarged unit cell, respectively, present obvious blue and red shifts; that is, the specific distances between the localization centers have been destroyed. Furthermore, the oscillation phase might change during the variation of state energy and modulation strength (figures 7.7(i)–(l)).

The distribution width of the LL wave functions could be tuned by enhancing the modulation strength, such as $B_M = B_z = 32$ T in figure 7.9. The widened oscillation modes are clearly revealed in the lowest conduction subband (the highest valence subband) among four non-degenerate subbands, as indicated from the wave functions of the $n_Q^c = 1$ energy subband at distinct wave vectors (the solid curves in figures 7.2(a)–(l)). This subband, with a very high dispersion relation near the band-edge state ($k_{be} = 1/3$) (figure 7.6(d)), presents the largest distribution width at k_{be} (figures 7.2(g) and (h)). Furthermore, for electronic states close to k_{be} (figures 7.2(e) and (f) and figures 7.2(i) and (j)), the symmetric/anti-symmetric distribution mode is changed into a highly asymmetric one. However, the zero-point number of the quantum oscillation modes is hardly affected by the strength of a modulated magnetic field.

The special structures in DOS could provide more information about the modulation effects. In general, each delta-function-like peak due to a uniform magnetic field (the dashed red curves in figure 7.10(a)) is changed into a pair of square-root asymmetric peaks, except that at E_F, as revealed under $B_z = 32$ T, $B_M = 4$ T and $R_M = R_B$ (the solid curves). Two neighboring peaks mainly arise from the band-edge states of the neighboring four non-degenerate energy subbands (figure 7.6(a)). Apparently, their energy spacing is the width of four separated subbands. The peak height is determined by the inverse of subband curvature. The drastic changes in energy spacing and peak intensity come to exist simultaneously in the increment of modulation strength, e.g. DOSs under $B_M = 8$ T and 16 T in figure 7.10(b). Furthermore, the number of peak structures is greatly reduced when the modulated and uniform components have the same strength ($B_M = B_z$ in figure 7.10(b) by the red curve). The main reason is that a lot of valley-like energy subbands (figure 7.6(d)) only make linear contributions to DOS, as clearly revealed at zero field (the black curve in figure 7.5(a)). However, the modulation period has weak effects on the main features of DOS, as clearly shown in figures 7.10(c) and (d). At $R < 1$, it could increase the band curvature and the number of band-edge states (figures 7.7(a) and (b)). The former and the latter make the peak structure weaker and stronger, respectively. The increase mostly compensates for the decrease, so that there are slight changes in the peak height (figure 7.10(c)). As for $R > 1$, the band curvature and range of k_x are reduced simultaneously (figures 7.7(c) and (d)). The competitive relationship between them also results in almost the same peak height (figure 7.10(d)). In short, the main features of LLs, energy dispersions, band-edge states, localization centers, distribution widths, symmetric/anti-symmetric modes,

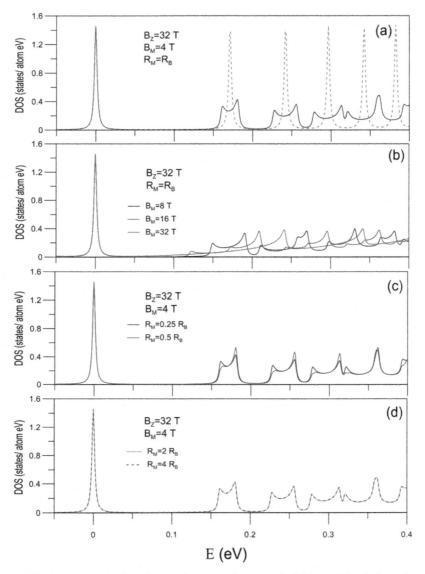

Figure 7.10. The low-energy density of states for composite magnetic fields modulated along the armchair direction for $B_z = 32$ and $R_M = R_B$ under (a) $B_M = 4$ T; (b) $B_M = 8$ T, 16 T and 32 T, and for $B_z = 32$ T and $B_M = 4$ T under (c) $R_M = R_B/4$ and $R_B/2$; (d) $R_M = 2R_B$ and $4R_B$. The red dashed curve in (a) represents the DOS in a uniform magnetic field.

oscillation phases, and special structures in DOS, could be changed by using a modulated magnetic field. A composite magnetic field is thus expected to create diverse essential properties, such as magneto-optical spectra and transport properties [1, 6].

References

[1] Ou Y C, Chiu Y H, Yang P H and Lin M F 2014 The selection rule of graphene in a composite magnetic field *Opt. Express* **22** 7473

[2] Chiu Y H, Lai Y H, Ho J H, Chuu D S and Lin M F 2008 Electronic structure of a two-dimensional graphene monolayer in a spatially modulated magnetic field: Peierls tight-binding model *Phys. Rev.* B **77** 045407

[3] Ho J H, Lai Y H, Chiu Y H and Lin M F 2008 Modulation effects on Landau levels in a monolayer graphen *Nanotechnology* **19** 035712

[4] Chiu Y H, Ho J H, Chang C P, Chuu D S and Lin M F 2008 Low-frequency magneto-optical excitations of a graphene monolayer: Peierls tight-binding model and gradient approximation calculation *Phys. Rev.* B **78** 245411

[5] Ajiki H and Ando T 1995 Lattice distortion of metallic carbon nanotubes induced by magnetic fields *J. Phys. Soc. Jpn.* **64** 260

[6] Park S and Sim H S 2008 Magnetic edge states in graphene in nonuniform magnetic fields *Phys. Rev.* B **77** 075433

[7] Ho J H, Chiu Y H, Tsai S J and Lin M F 2009 Semimetallic graphene in a modulated electric potential *Phys. Rev.* B **79** 115427

[8] Matulis A and Peeters F M 2007 Appearance of enhanced Weiss oscillations in graphene: theory *Phys. Rev.* B **75** 125429

[9] Tahir M and Sabeeh K 2007 Theory of Weiss oscillations in the magnetoplasmon spectrum of Dirac electrons in graphene *Phys. Rev.* B **76** 195416

[10] Park C-H, Yang L, Son Y-W, Cohen M L and Louie S G 2008 Anisotropic behaviours of massless Dirac fermions in graphene under periodic potentials *Nature Phys.* **4** 213

[11] Park C-H, Yang L, Son Y-W, Cohen M L and Louie S G 2008 New generation of massless Dirac fermions in graphene under external periodic potentials *Phys. Rev. Lett.* **101** 126804

[12] Ou Y C, Sheu J K, Chiu Y H, Chen R B and Lin M F 2011 Influence of modulated fields on the Landau level properties of graphene *Phys. Rev.* B **83** 195405

Chapter 8

Concluding remarks

In this work, the generalized tight-binding model, based on the subenvelope functions of distinct sublattices, is developed to explore the feature-rich magneto-electronic properties of layered systems. Such a model is suitable for various symmetric lattices [1–15], multi-layer structures [5, 7–10], low-symmetry stacking configurations [5, 10], distinct dimensions [1–12], multi-orbital bondings, coupling interactions of orbital and spin, composite external fields [16], and uniform and modulated fields [17, 18]. It is useful in understanding the essential physical properties, e.g. the diverse magneto-optical selection rules [5, 11, 12, 15–17, 19–25] and the LL-induced plasmons [3, 26, 27]. Moreover, this method could be further used to solve new Hamiltonians of emerging materials under the influence of external fields.

Layered systems exhibit unusual energy bands and rich LLs in terms of the spatial distributions, orbital components, spin configurations, state degeneracy, and external-field dependences. Well-behaved, perturbed and undefined LLs are revealed in sliding graphenes, especially for the third ones mainly coming from the dramatic transformation between two high-symmetry stacking configurations [5]. For ABC- and AAB-stacked graphenes, the complicated interlayer hopping integrals result in abnormal $n^{c,v}$ ordering and non-monotonic B_z-dependence [7, 10]. Intragroup and intergroup LL anti-crossings occur frequently, clearly illustrating the strong competition of the distinct constant-energy loops in the magnetic quantization. The SOC can create up- and down-dominated LLs in silicene, germanene, tinene and MoS_2. Tinene and MoS_2 have low-lying LLs composed of different orbitals and spin configurations, mainly owing to the cooperation of the critical multi-orbital bondings and SOC. Concerning few-layer phosphorenes, the puckered structure induces intralayer and interlayer multi-hopping integrals and thus a special dependence of LL energies on $(n^{c,v}, B_z)$. The LL state degeneracy is reduced when the inversion symmetry of $z \rightarrow -z$ ($x \rightarrow -x$) is destroyed, or two equivalent valleys are absent. For example, there are four-fold degenerate LLs in AAB-stacked graphene [10] and few-layer phosphorenes, and doubly degenerate $n_\Gamma^{c,v}$ LLs in tinene.

Specifically, MoS_2 exhibits spin- and valley-dependent LL subgroups, which arise from the SOC and the site-energy differences in the B_z-enlarged unit cell, respectively. The LL splittings are easily observed in the presence of a perpendicular electric field except for AA-stacked graphenes and monolayer silicene. Moreover, they can induce frequent anti-crossings and crossings in the V_z-dependent energy spectra. The two anti-crossing LLs, which are, respectively, associated with the specific interlayer hopping integrals (layered graphenes), the significant SOCs (germanene and tinene), and the complicated intralayer and interlayer hopping integrals (bilayer phosphorene), have quantum number differences: $\Delta n = 3i$, ± 1 and i. As for bilayer phosphorene, these two LLs come from distinct subgroups initiated from the Dirac and Γ points. The above-mentioned diverse LL energy spectra are directly reflected in the structure, height, energy and number of the prominent DOS peaks; furthermore, they can be verified by STS measurements.

Non-uniform magnetic quantization presents unusual electronic properties in monolayer graphene. A spatially modulated magnetic field can create a lot of QLLs and thus lead to a reduction in dimensionality, the coexistence of weak and strong energy dispersions, anisotropic band structure, composite behavior in state degeneracy, extra band-edge states, and asymmetry of energy bands about k_y^o. The OLLs near the original band-edge state are similar to LLs in the main characteristics. However, significant differences come to exist for the other states, such as the separation/combination of two localization centers, the regular/irregular oscillation mode and the probability transfer. The van Hove singularities in DOS behave like the 1D square-root asymmetric peaks, except for the delta-function-like prominent peak at E_F. Moreover, this field can dramatically change the magneto-electronic properties of LLs. The dispersionless LLs become a lot of 1D parabolic/valley-like subbands, in which energy dispersions and band-edge state energies strongly depend on the strength of a modulated magnetic field. Such subbands could be regarded as QLLs, according to the variations in the localization center, width, phase and symmetric/anti-symmetric distribution of the spatial oscillation modes. DOS in a composite magnetic field especially presents many pair-like asymmetric peaks being sensitive to the modulation strength, but not the period.

Also, the different dimensions can diversify the magnetic quantization. 3D graphites and 1D graphene nanoribbons are very different from 2D graphenes in the quantized electronic properties. As to graphites, the periodical interlayer hopping integrals induce Landau subbands with energy dispersions along $\hat{k}_z$. AA-, AB- and ABC-stacked graphites, respectively, possess one group, two groups, and one group of valence and conduction Landau subbands, in which the band widths are about 1 eV, 0.2 eV and 0.01 eV [1, 2, 4, 23, 28]. However, there are N groups of LLs in N-layer graphene systems. AA- and ABC-stacked graphites exhibit monolayer-like wave functions, while AB-stacked graphite displays monolayer- and bilayer-like spatial distributions. In sharp contrast to the ABC-stacked graphenes, the anti-crossing phenomena in the B_z-dependent energy spectrum are absent in rhombohedral graphite. However, the magneto-electronic properties of graphene nanoribbons are mainly determined by the rather strong competition between the magnetic quantization and the finite-width lateral confinement [11]. When ribbon widths are larger than magnetic lengths, 1D nanoribbons have many composite energy subbands. Each subband is composed of a dispersionless QLL and parabolic dispersions along $\hat{k}_x$. Such QLLs belong to the well-behaved modes localized at the ribbon center. Their magneto-optical selection rule is similar to that of monolayer graphene [12]. It should be noted that the QLLs in 1D graphene nanoribbons are very different from those in 2D monolayer graphene. The former and the latter, respectively, arise from the quantum confinement and the magnetic modulation effect, corresponding to the open and periodic boundary conditions. The important differences also cover the existence of quantization modes, state degeneracy, probability distributions, and number of localization centers. In addition, a carbon nanotube could be regarded as a rolled-up graphene tubule. Except for a super-high magnetic field ($B_z > 10^5$T; [29]), it is impossible to generate quantized QLLs in a curved cylinder because of the vanishing net magnetic flux.

The generalized tight-binding model can combine with single- and many-particle theories to explore other essential physical properties in detail, since it provides reliable band structures, LL energy spectra, and wave functions under the various external fields. Combinations with the static Kubo formula, the frequency-dependent one and the layer-dominated RPA, are very suitable in studying the Hall effect (QHE), optical and magneto-optical properties, and the Coulomb excitations, respectively. A recent study on QHE shows that this model can determine the available inter-LL transitions of the electrical conductivity; i.e. the static selection rules are obtained exactly even in the presence of complicated magneto-electronic structures [30]. Three kinds of LLs, the crossing, anti-crossing and splitting energy spectra are predicted to create unusual transport properties, covering the non-integer QHE, integer conductivities with different heights, splitting-induced reduction and complexity of quantum conductivity, a vanishing or finite conductivity at the neutral point, and the staircase, well-like, composite, and abnormal plateau structures during the variation of the magnetic field. Similar studies could be further generalized to the emergent 2D systems in the observation of more quantum transport phenomena.

The direct combination of the generalized tight-binding model and the frequency-dependent Kubo formula is very useful in a full understanding of optical properties in the presence/absence of magnetic and electric fields. The electric dipole moment, which determines the absorption intensity of the available vertical excitation channels, could be evaluated from the layer-dependent subenvelope functions under any external fields. Up to now, the stacking-, layer-number-, dimension-, and field-enriched absorption spectra have been studied for few-layer graphenes [5, 21, 24, 25], graphites [22, 23], carbon nanotubes [31], and graphene nanoribbons [11, 12, 20]. Specifically, complete studies on AA-, AB-, ABC- and AAB-stacked graphenes and sliding bilayer systems present the rich and unique optical spectra, mainly owing to the complex relations among the interlayer atomic interactions, magnetic quantization and Coulomb potential energy [5]. The zero-magnetic-field absorption spectra of 2D systems have three types of structures, shoulders, asymmetric peaks and logarithmic peaks, reflecting the band-edge states of parabolic dispersions, constant-energy loops and saddle points, respectively [32, 33]. The low-frequency threshold excitations are forbidden only in even-layer AA stacking systems without the Dirac point at the Fermi level [24, 34]; furthermore, optical gaps and special structures could be generated by an electric field [35]. A lot of delta-function-like symmetric peaks, as shown in DOS, are clearly revealed in magneto-optical absorption spectra. The single, twin and double peaks, respectively, come from the symmetric, asymmetric and splitting LL energy spectra [21, 36]. The distinct magneto-optical selection rules are induced by three kinds of LLs [5]. AAB stacking, without the mirror symmetry, possesses more absorption structures compared with the other stackings [37]. Moreover, the frequent LL anti-crossings in the magnetic- and electric-field-dependences lead to extra absorption peaks and lower intensities. The unusual LLs in silicene, germanene, tinene, MoS_2 and phosphorene, as discussed in chapters 4–6, are expected to exhibit diversified magneto-optical properties. Their excitation phenomena are worthy of thorough investigation.

As for Coulomb excitations closely related to band structures, a delicate theory within RPA has been successfully developed for 2D graphene systems [26, 27, 38, 39]. The theoretical calculations are based on the layer-dependent subenvelope functions, being identical to those used in the generalized tight-binding model. Consistent combination of the two models can enable a full investigation of the diverse Coulomb excitation behaviors even induced by the various external fields [26, 27, 40–42]. The most important characteristic of this combination lies in the simultaneous consider-ations of the intralayer and interlayer hopping integrals, the intralayer and interlayer Coulomb interactions, and the magnetic and electric fields. Such a method is very different from the low-energy perturbation approximation. Apparently, few-layer graphenes, silicene and germanene can exhibit rich and unique single- and many-particle excitations, in which their main features, the momentum- and energy-depend-ences of electron–hole pairs and plasmon modes, strongly depend on the band structures and LLs. Theoretical calculations have been made for monolayer graphene in a B_z-field [27], bilayer AA and AB stackings without/with magnetic quantization [26, 38, 39], few-layer graphenes under a perpendicular electric field [40–42], silicene in the presence/absence of external fields [43–45], and germanene [46]. In particular, diverse phase diagrams associated with transferred momenta and energies exist. Distinct plasmons (magneto-plasmons) and complex Landau damping are absent in 2D electron gas systems. The Coulomb excitations and decays in emergent 2D materials should be one of the mainstream topics in future systematic studies. In summary, the generalized tight-binding model can efficiently deal with the essential physical proper-ties of condensed-matter systems in terms of energy spectra, wave functions, physical pictures and direct combinations with the other theories.

Acknowledgment

This work was supported by the NSC of Taiwan, under Grant No. NSC 105-2112-M-006-002-MY3. We thank Godfrey Gumbs for critical comments on this book.

References

[1] Ho Y H, Wang J, Chiu Y H, Lin M F and Su W P 2011 Characterization of Landau subbands in graphite: a tight-binding study *Phys. Rev.* B **83** 121201

[2] Ho C H, Ho Y H, Liao Y Y, Chiu Y H, Chang C P and Lin M F 2012 Diagonalization of Landau level spectra in rhombohedral graphite *J. Phys. Soc. Jpn.* **81** 024701

[3] Chen R B, Chiu C W and Lin M F 2015 Magnetoplasmons in simple hexagonal graphite *RSC Adv* **5** 53736–40

[4] Ho C H, Chang C P and Lin M F 2014 Landau subband wave functions and chirality manifestation in rhombohedral graphite *Solid State Commun.* **197** 11–5

[5] Huang Y K, Chen S C, Ho Y H, Lin C Y and Lin M F 2014 Feature-rich magnetic quantization in sliding bilayer graphenes *Sci. Rep.* **4** 7509

[6] Ho J H, Lai Y H, Chiu Y H and Lin M F 2008 Landau levels in graphene *Physica* E **40** 1722–5

[7] Lin C Y, Wu J Y, Chiu Y H and Lin M F 2014 Stacking-dependent magneto-electronic properties in multilayer graphenes *Phys. Rev.* B **90** 205434

[8] Lai Y H, Ho J H, Chang C P and Lin M F 2008 Magnetoelectronic properties of bilayer bernal graphene *Phys. Rev.* B **77** 085426

[9] Lin C Y, Wu J Y, Ou Y J, Chiu Y H and Lin M F 2015 Magneto-electronic properties of multilayer graphenes *Phys. Chem. Chem. Phys.* **17** 26008–35

[10] Do T N, Lin C Y, Lin Y P, Shih P H and Lin M F 2015 Configuration-enriched magnetoelectronic spectra of AAB-stacked trilayer graphene *Carbon* **94** 619–32

[11] Huang Y C, Chang C P and Lin M F 2007 Magnetic and quantum confinement effects on electronic and optical properties of graphene ribbons *Nanotechnology* **18** 495401

[12] Chung H C, Chang C P, Lin C Y and Lin M F 2016 Electronic and optical properties of graphene nanoribbons in external fields *Phys. Chem. Chem. Phys.* **18** 7573–616

[13] Chen S C, Wu C L, Wu J Y and Lin M F 2016 Magnetic quantization of sp^3 bonding in monolayer gray tin *Phys. Rev.* B **94** 045410

[14] Ho Y H, Su W P and Lin M F 2015 Hofstadter spectra for d-orbital electrons: a case study on MoS_2 *RSC Adv* **5** 20858–64

[15] Ho Y H, Wang Y H and Chen H Y 2014 Magnetoelectronic and optical properties of a MoS^2 monolayer *Phys. Rev* B **89** 55316

[16] Ou Y C, Chiu Y H, Yang P H and Lin M F 2014 The selection rule of graphene in a composite magnetic field *Opt. Express* **22** 7473

[17] Ou Y C, Chiu Y H, Lu J M, Su W P and Lin M F 2013 Electric modulation effect on magneto-optical spectrum of monolayer graphene *Comput. Phys. Commun.* **184** 1821–6

[18] Ou Y C, Sheu J K, Chiu Y H, Chen R B and Lin M F 2011 Influence of modulated fields on the Landau level properties of graphene *Phys. Rev.* B **83** 195405

[19] Ho Y H, Chiu Y H, Lin D H, Chang C P and Lin M F 2010 Magneto-optical selection rules in bilayer Bernal graphene *ACS Nano* **4** 1465–72

[20] Huang Y C, Chang C P and Lin M F 2008 Magnetoabsorption spectra of bilayer graphene ribbons with Bernal stacking *Phys. Rev.* B **78** 115422

[21] Lin Y P, Lin C Y, Ho Y H, Do T N and Lin M F 2015 Magneto-optical properties of ABC-stacked trilayer graphene *Phys. Chem. Chem. Phys.* **17** 15921–7

[22] Chen R B, Chiu Y H and Lin M F 2014 Beating oscillations of magneto-optical spectra in simple hexagonal graphite *Comput. Phys. Commun.* **189** 60–5

[23] Chen R B, Chiu Y H and Lin M F 2012 A theoretical evaluation of the magneto-optical properties of AA-stacked graphite *Carbon* **54** 268–76

[24] Ho Y H, Wu J Y, Chen R B, Chiu Y H and Lin M F 2010 Optical transitions between Landau levels: AA-stacked bilayer graphene *Appl. Phys. Lett.* **97** 101905

[25] Ho Y H, Chiu C W, Su W P and Lin M F 2014 Magneto-optical spectra of transition metal dichalcogenides: a comparative study *Appl. Phys. Lett.* **105** 222411

[26] Wu J Y, Gumbs G and Lin M F 2014 Combined effect of stacking and magnetic field on plasmon excitations in bilayer graphene *Phys. Rev.* B **89** 165407

[27] Wu J Y, Chen S C, Roslyak O, Gumbs G and Lin M F 2011 Plasma excitations in graphene: their spectral intensity and temperature dependence in magnetic field *ACS Nano* **5** 1026–32

[28] Ho C H, Chang C P, Su W P and Lin M F 2013 Precessing anisotropic dirac cone and Landau subbands along a nodal spiral *New J. Phys.* **15** 053032

[29] Ajiki H and Ando T 1996 Energy bands of carbon nanotubes in magnetic fields *J Phys. Soc. Jpn.* **65** 505

[30] Do T N, Chang C P, Shih P H and Lin M F 2017 Stacking-enriched magnetotransport properties of few-layer graphenes arXiv:1704.01313

[31] Shyu F L, Chang C P, Chen R B, Chiu C W and Lin M F 2003 Magnetoelectronic and optical properties of carbon nanotubes *Phys. Rev.* B **67** 045405

[32] Lu C L, Chang C P, Huang Y C, Chen R B and Lin M F 2006 Influence of an electric field on the optical properties of few-layer graphene with AB stacking *Phys. Rev.* B **73** 144427

[33] Lu C L, Lin H L, Hwang C C, Wang J, Chang C P and Lin M F 2006 Absorption spectra of tri-layer rhombohedral graphite *Appl. Phys. Lett.* **89** 221910

[34] Chiu C W, Chen S C, Huang Y C, Shyu F L and Lin M F 2013 Critical optical properties of AA-stacked multilayer graphenes *Appl. Phys. Lett.* **103** 041907

[35] Tsai S J, Chiu Y H, Ho Y H and Lin M F 2012 Gate-voltage-dependent Landau levels in AA-stacked bilayer graphene *Chem. Phys. Lett.* **550** 104–10

[36] Lin Y P, Lin C Y, Chang C P and Lin M F 2015 Electric-field-induced rich magneto-absorption spectra of ABC-stacked trilayer graphene *RSC Adv* **5** 80410

[37] Do T N, Shih P H, Chang C P, Lin C Y and Lin M F 2016 Rich magneto-absorption spectra in AAB-stacked trilayer graphene *Phys. Chem. Chem. Phys.* **18** 17597

[38] Ho Y H, Chang C P and Lin M F 2006 Electronic excitations of the multilayered graphite *Phys. Lett.* A **352** 446–50

[39] Ho Y H, Lu C L, Hwang C C, Chang C P and Lin M F 2006 Coulomb excitations in AA- and AB-stacked bilayer graphites *Phys. Rev.* B **74** 085406

[40] Lin M F, Chuang Y C and Wu J Y 2012 Electrically tunable plasma excitations in AA-stacked multilayer graphene *Phys. Rev* B **86** 125434

[41] Chuang Y C, Wu J Y and Lin M F 2013 Electric field dependence of excitation spectra in AB-stacked bilayer graphene *Sci. Rep.* **3** 1368

[42] Chuang Y C, Wu J Y and Lin M F 2013 Electric-field-induced plasmon in AA-stacked bilayer graphene *Ann. Phys.* **339** 298

[43] Wu J Y, Lin C Y, Gumbs G and Lin M F 2015 The effect of perpendicular electric field on temperature-induced plasmon excitations for intrinsic silicene *RSC Adv.* **5** 51912–8

[44] Wu J Y, Chen S C and Lin M F 2014 Temperature-dependent coulomb excitations in silicene *New J. Phys.* **16** 125002

[45] Wu J Y, Chen S C, Gumbs G and Lin M F 2016 Feature-rich electronic excitations in external fields of 2D silicene *Phys. Rev.* B **94** 205427

[46] Shih P H, Chiu Y H, Wu J Y, Shyu F L and Lin M F 2017 Coulomb excitations of monolayer germanene *Sci. Rep.* **7** 40600